T0325910

lectures on
quantum mechanics
basic matters

lectures on
quantum mechanics

basic matters

Berthold-Georg Englert
National University of Singapore, Singapore

World Scientific

NEW JERSEY · LONDON · SINGAPORE · BEIJING · SHANGHAI · HONG KONG · TAIPEI · CHENNAI

Published by

World Scientific Publishing Co. Pte. Ltd.

5 Toh Tuck Link, Singapore 596224

USA office: 27 Warren Street, Suite 401-402, Hackensack, NJ 07601

UK office: 57 Shelton Street, Covent Garden, London WC2H 9HE

British Library Cataloguing-in-Publication Data
A catalogue record for this book is available from the British Library.

LECTURES ON QUANTUM MECHANICS
(In 3 Volumes)
Volume 1: Basic Matters

Copyright © 2006 by World Scientific Publishing Co. Pte. Ltd.

For photocopying of material in this volume, please pay a copying fee through the Copyright Clearance Center, Inc., 222 Rosewood Drive, Danvers, MA 01923, USA. In this case permission to photocopy is not required from the publisher.

ISBN-13 978-981-256-790-1 (Set)
ISBN-10 981-256-790-9   (Set)
ISBN-13 978-981-256-791-8 (pbk) (Set)
ISBN-10 981-256-791-7   (pbk) (Set)

ISBN-13 978-981-256-970-7 (Vol. 1)
ISBN-10 981-256-970-7   (Vol. 1)
ISBN-13 978-981-256-971-4 (pbk) (Vol. 1)
ISBN-10 981-256-971-5   (pbk) (Vol. 1)

Printed in Singapore

To my teachers, colleagues, and students

Preface

This book on the *Basic Matters* of quantum mechanics grew out of a set of lecture notes for a second-year undergraduate course at the National University of Singapore (NUS). It is a first introduction that does not assume any prior knowledge of the subject. The presentation is rather detailed and does not skip intermediate steps that — as experience shows — are not so obvious for the learning student.

Starting from the simplest quantum phenomenon, the Stern–Gerlach experiment with its choice between two discrete outcomes, and ending with the standard examples of one-dimensional continuous systems, the physical concepts and notions as well as the mathematical formalism of quantum mechanics are developed in successive small steps, with scores of exercises along the way. The presentation is "modern", a dangerous word, in the sense that the natural language of the trade — Dirac's kets and bras and all that — is introduced early, and the temporal evolution is dealt with in a picture-free manner, with Schrödinger's and Heisenberg's equations of motion side by side and on equal footing.

Two companion books on *Simple Systems* and *Perturbed Evolution* cover the material of the subsequent courses at NUS for third- and fourth-year students, respectively. The three books are, however, not strictly sequential but rather independent of each other and largely self-contained. In fact, there is quite some overlap and a considerable amount of repeated material. While the repetitions send a useful message to the self-studying reader about what is more important and what is less, one could do without them and teach most of *Basic Matters*, *Simple Systems*, and *Perturbed Evolution* in a coherent two-semester course on quantum mechanics.

All three books owe their existence to the outstanding teachers, colleagues, and students from whom I learned so much. I dedicate these lectures to them.

I am grateful for the encouragement of Professors Choo Hiap Oh and Kok Khoo Phua who initiated this project. The professional help by the staff of World Scientific Publishing Co. was crucial for the completion; I acknowledge the invaluable support of Miss Ying Oi Chiew and Miss Lai Fun Kwong with particular gratitude. But nothing would have come about, were it not for the initiative and devotion of Miss Jia Li Goh who turned the original handwritten notes into electronic files that I could then edit.

I wish to thank my dear wife Ola for her continuing understanding and patience by which she is giving me the peace of mind that is the source of all achievements.

Singapore, March 2006   *BG Englert*

Contents

Preface   vii

1.   A Brutal Fact of Life   1

    1.1 Causality and determinism   1
    1.2 Bell's inequality: No hidden determinism   4
    1.3 Remarks on terminology   8

2.   Kinematics: How Quantum Systems are Described   11

    2.1 Stern–Gerlach experiment   11
    2.2 Successive Stern–Gerlach measurements   14
    2.3 Order matters   17
    2.4 Mathematization   17
    2.5 Probabilities and probability amplitudes   23
    2.6 Quantum Zeno effect   32
    2.7 Kets and bras   35
    2.8 Brackets, bra-kets, and ket-bras   38
    2.9 Pauli operators, Pauli matrices   41
    2.10 Functions of Pauli operators   44
    2.11 Eigenvalues, eigenkets, eigenbras   46
    2.12 Wave-particle duality   51
    2.13 Expectation value   51
    2.14 Trace   54
    2.15 Statistical operator   56
    2.16 Mixtures and blends   60
    2.17 Nonselective measurement   61
    2.18 Entangled atom pairs   63

2.19 State reduction 68
2.20 Measurements with more than two outcomes 70
2.21 Unitary operators 75
2.22 Hermitian operators 78

3. Dynamics: How Quantum Systems Evolve   81

   3.1 Schrödinger equation 81
   3.2 Heisenberg equation 85
   3.3 Equivalent Hamilton operators 87
   3.4 von Neumann equation 88
   3.5 Example: Larmor precession 89
   3.6 Time-dependent probability amplitudes 93
   3.7 Schrödinger equation for probability amplitudes 94
   3.8 Time-independent Schrödinger equation 98
   3.9 Example: Two magnetic silver atoms 101

4. Motion along the x Axis   109

   4.1 Kets, bras, wave functions 109
   4.2 Position operator 114
   4.3 Momentum operator 117
   4.4 Heisenberg's commutation relation 119
   4.5 Position-momentum transformation function 119
   4.6 Expectation values 122
   4.7 Uncertainty relation 125
   4.8 State of minimum uncertainty 129
   4.9 Time dependence 132
   4.10 Excursion into classical mechanics 132
   4.11 Hamilton operator, Schrödinger equation 136
   4.12 Time transformation function 139

5. Elementary Examples   141

   5.1 Force-free motion 141
      5.1.1 Time-transformation functions 141
      5.1.2 "Spreading" of the wave function 143
      5.1.3 Long-time and short-time behavior 149
      5.1.4 Interlude: General position-dependent force 155
      5.1.5 Energy eigenstates 157
   5.2 Constant force 160

    5.2.1   Energy eigenstates   160
    5.2.2   Limit of no force   162
5.3   Harmonic oscillator   165
    5.3.1   Energy eigenstates: Power-series method   165
    5.3.2   Energy eigenstates: Ladder-operator approach   170
    5.3.3   Hermite polynomials   176
    5.3.4   Infinite matrices   177
5.4   Delta potential   181
    5.4.1   Bound state   181
    5.4.2   Scattering states   186
5.5   Square-well potential   191
    5.5.1   Bound states   191
    5.5.2   Delta potential as a limit   196
    5.5.3   Scattering states. Tunneling   197

Index   203

Chapter 1

A Brutal Fact of Life

1.1 Causality and determinism

Before their first encounter with the quantum phenomena that govern the realm of atomic physics and sub-atomic physics, students receive a training in classical physics, where Isaac Newton's mechanics of massive bodies and James C. Maxwell's electromagnetism — the physical theory of the electromagnetic field and its relation to the electric charges — give convincingly accurate accounts of the observed phenomena. Indeed, almost all experiences of physical phenomena that we are conscious of without the help of refined instruments fit perfectly into the conceptual and technical framework of these classical theories. It is instructive to recall two characteristic features that are equally possessed by Newton's mechanics and Maxwell's electromagnetism: *Causality* and *Determinism*.

Causality is inference in time: Once you know the state of affairs — physicists prefer to speak more precisely of the "state of the system" — you can predict the state of affairs at any later time, and often also retrodict the state of affairs at earlier times. Having determined the relative positions of the sun, earth and moon and their relative velocities, we can calculate highly precisely when the next lunar eclipse will happen (extreme precision on short time scales and satisfactory precision for long time scales also require good knowledge of the positions and velocities of the other planets and their satellites, but that is a side issue here) or when the last one occurred. Quite similarly, present knowledge of the strength and direction of the electric and magnetic fields together with knowledge about the motion of the electric charges enables us to calculate reliably the electromagnetic field configuration in the future, or the past.

Causality, as we shall see, is also a property of quantal evolution: Given the state of the system now, we can infer the state of the system later (but, typically, not earlier). Such as there are Newton's equation of motion in mechanics, and Maxwell's set of equations for the electromagnetic field, there are also equations of motion in quantum mechanics: Erwin Schrödinger's equation, which is more in the spirit of Maxwell's equations, and Werner Heisenberg's equation, which is more in Newton's tradition.

We say that the classical theories are *deterministic* because the state of the system uniquely determines all phenomena. When the positions and velocities of all objects are known in Newton's mechanics, also the results of all possible measurements are predictable, there is no room for any uncertainty in principle. Likewise, once the electromagnetic field is completely specified and the positions and velocities of all charges are known in Maxwell's electromagnetic theory, all possible electromagnetic phenomena are fully predictable.

Let us look at a somewhat familiar situation that illustrates this point and will enable us to establish the difference in situation that we encounter in quantum physics. You have all seen reflections of yourself in the glass of a shopping window, while at the same time having a good view of the goodies for sale. This is a result of the property of the glass sheet that it partly transmits light and partly reflects it. In a laboratory version we could have 50% probability each for transmission and reflection:

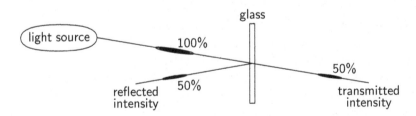

A light source emits pulses of light, which are split in two by such a half-transparent mirror, half of the intensity being transmitted, the other half reflected. Given the properties of pulses emitted by the source and the material properties of the glass, we can predict completely how much of the intensity is reflected, how much is transmitted, how the pulse shape is changed, and so forth — all these being implications of Maxwell's equations.

But, we know that there is a different class of phenomena that reveal a certain graininess of light: the pulses consist of individual lumps of energy — "light quanta", or "photons". (We are a bit sloppy with the terminology here, at a more refined level, photons and light quanta are not the same, but that is irrelevant presently.) We become aware of the photons if we dim the light source by so much that there is only a single photon per pulse. We also register the reflected and transmitted light by single-photon counters:

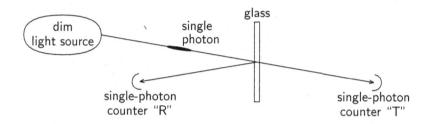

What will be the fate of the next photon to come? Since it cannot split in two, either the photon is transmitted as a whole, or it is reflected as a whole, so that eventually *one* of the counters will register the photon. A single photon, so to say, makes one detector click: either we register a click of detector T or of detector R, but not of both.

What is important here is that we *cannot predict* which detector will click for the next photon, all we know is the history of the clicks of the photons that have already arrived. Perhaps a sequence such as

R R T R T T T R T R

was the case for the last ten photons. In a long sequence, reporting the detector clicks of very many photons, there will be about the same number of T clicks and R clicks, because it remains true that half of the intensity is reflected and half transmitted. On the single-photon level, this becomes a *probabilistic* fact: Each photon has a 50% chance of being reflected and an equal chance of being transmitted. And this is *all* we can say about the future fate of a photon approaching the glass sheet.

So, when repeating the experiment with another set of ten photons, we do not reproduce the above sequence of detector clicks, but rather get another one, perhaps

R T T R R R R T R T

And a third set of ten would give yet another sequence, all 2^{10} possible sequences occurring with the same frequency if we repeat the experiment very often.

Thus, although we know exactly all the properties of the incoming photon, we cannot predict which detector will click. We can only make statistical predictions that answer questions such as "How likely are four Ts and six Rs in the next sequence of ten?"

1-1   Answer this question.

What we face here, in a simple but typical situation, is the *lack of determinism* of quantum phenomena. Complete knowledge of the state of affairs does not enable us to predict the outcomes of all measurements that could be performed on the system. In other words, the state does not determine the phenomena. There is a fundamental element of chance: The laws of nature that connect our knowledge about the state of the system with the observed phenomena are *probabilistic*, not deterministic.

1.2   Bell's inequality: No hidden determinism

Now, that raises the question about the origin of this probabilistic nature. Does the lack of determinism result from a lack of knowledge? Or, put differently, could we know more than we do, and then have determinism reinstalled? The answer is *No*. Even if we know everything that can possibly be known about the photon, we cannot predict its fate.

It is not simple to make this point for the example discussed above with simple photons incident on a half-transparent mirror. In fact, one can construct contrived formalisms in which the photons are equipped with internal clockworks of some sort that determine in a hidden fashion where each photon will go. But in more complicated situations, even the most ingenious deterministic mechanism cannot reproduce the observed facts in all respects. The following argument is a variant of the one given by John S. Bell in the 1960s.

Consider the more general scenario in which a photon-pair source always emits two photons, one going to the left, the other going to the right:

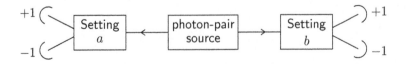

Each photon is detected by one of two detectors eventually — with measurement results $+1$ or -1 — and the devices allow for a number of parameter settings. We denote by symbol a the collection of parameters on the left, and by b those on the right. Details do not matter, all we need is that different settings are possible, that is: there is a choice between different measurements on both sides. The only restriction we insist upon is that there are only two possible outcomes for each setting, the abstract generalization of "transmission" and "reflection" in the single-photon plus glass sheet example above.

For any given setting, the experimental data is of this kind:

photon pair no.	1	2	3	4	5	6	7	8	...
on the left	$+1$	$+1$	-1	-1	$+1$	-1	$+1$	$+1$	...
on the right	$+1$	-1	-1	$+1$	-1	-1	-1	$+1$	...
product	$+1$	-1	$+1$	-1	-1	$+1$	-1	$+1$	...

The products in the last row distinguish the pairs with the same outcomes on the left and the right (product $= +1$) from those with opposite outcomes (product $= -1$). We use these products to define the *Bell correlation* $C(a,b)$ for the chosen setting specified by parameters a, b,

$$C(a,b) = \frac{(\text{number of } +1 \text{ pairs}) - (\text{number of } -1 \text{ pairs})}{\text{total number of observed pairs}}. \qquad (1.2.1)$$

Clearly, we have $C(a,b) = +1$ if $+1$ on one side is always matched with a $+1$ on the other and -1 with -1, and we have $C(a,b) = -1$ if $+1$ is always paired with -1 and -1 with $+1$. In all other cases, the value of $C(a,b)$ is between these extrema, so that

$$-1 \leq C(a,b) \leq +1 \qquad (1.2.2)$$

for any setting a, b.

Following Bell, let us now fantasize about a (hidden) mechanism that determines the outcome on each side. We conceive each pair as being characterizable by a set of parameters collectively called λ, and that the source realizes various λ with different relative frequencies. Thus, there is a positive weight function $\rho(\lambda)$, such that $d\lambda\,\rho(\lambda)$ is the probability of having a λ value within a $d\lambda$ volume around λ. These probabilities must be positive numbers that sum up to unity,

$$\rho(\lambda) \geq 0, \qquad \int d\lambda\,\rho(\lambda) = 1. \qquad (1.2.3)$$

We need not be more specific because further details are irrelevant to the argument — which is, of course, the beauty of it.

We denote by $A_\lambda(a)$ the measurement result on the left for setting a when the hidden control parameter has value λ, and by $B_\lambda(b)$ the corresponding measurement result on the right. Since all measurement results are either $+1$ or -1, we have

$$A_\lambda(a) = \pm 1, \qquad B_\lambda(b) = \pm 1 \qquad \text{for all} \quad a, b, \lambda \qquad (1.2.4)$$

and also

$$A_\lambda(a)B_\lambda(b) = \pm 1 \qquad \text{for all} \quad a, b, \lambda. \qquad (1.2.5)$$

This is then the product to be entered in the table above, before (1.2.1), for the pair that leaves the source with value λ and encounters the settings a and b. Upon summing over all pairs, we get

$$C(a, b) = \int \mathrm{d}\lambda\, \rho(\lambda) A_\lambda(a) B_\lambda(b) \qquad (1.2.6)$$

for the Bell correlation, and all the rest follows from this expression.

Before proceeding, however, let us note that an important assumption has entered: We take for granted that the measurement result on the left does not depend on the setting of the apparatus on the right, and vice versa. This is an expression of *locality* as we naturally accept it as a consequence of Albert Einstein's observation that spatially well separated events cannot be connected by any causal links if they are simultaneous in one reference frame. Put differently, if the settings a and b are decided very late, just before the measurements actually take place, any influence of the setting on one side upon the outcome on the other side would be inconsistent with Einsteinian causality. With this justification there is no need to consider the more general possibility of having $A_\lambda(a, b)$ on the left and $B_\lambda(a, b)$ on the right. Such a b dependence of A_λ and an a dependence of B_λ are physically unacceptable, but of course it remains a mathematical possibility that cannot be excluded on purely logical grounds.

All together we now consider two settings on the left, a and a', and two on the right, b and b'. The difference between the Bell correlations for settings a, b and a, b' is then

$$
\begin{aligned}
C(a,b) - C(a,b') &= \int d\lambda\, \rho(\lambda)[A_\lambda(a)B_\lambda(b) - A_\lambda(a)B_\lambda(b')] \\
&= \int d\lambda\, \rho(\lambda)A_\lambda(a)B_\lambda(b)[1 \pm A_\lambda(a')B_\lambda(b')] \\
&\quad - \int d\lambda\, \rho(\lambda)A_\lambda(a)B_\lambda(b')[1 \pm A_\lambda(a')B_\lambda(b)] \quad (1.2.7)
\end{aligned}
$$

where the "$\pm$" terms compensate for each other provided we take either both upper signs or both lower signs. Now, since

$$
\rho(\lambda) \geq 0, \qquad |A_\lambda(a)B_\lambda(b)| = 1, \qquad 1 \pm A_\lambda(a)B_\lambda(b) \geq 0 \qquad (1.2.8)
$$

for both signs and all values of λ, a, b, repeated applications of the triangle inequality gives

$$
\begin{aligned}
|C(a,b) - C(a,b')| &\leq \int d\lambda\, \rho(\lambda)[1 \pm A(a')B(b)] \\
&\quad + \int d\lambda\, \rho(\lambda)[1 \pm A(a')B(b')] \\
&= 2 \pm [C(a',b) + C(a',b')]. \qquad (1.2.9)
\end{aligned}
$$

Consequently, the left-hand side cannot exceed the smaller one of the two right-hand sides (one for $+$ and one for $-$ in $\pm$),

$$
|C(a,b) - C(a,b')| \leq 2 - |C(a',b) + C(a',b')| \qquad (1.2.10)
$$

or, after rearranging,

$$
|C(a,b) - C(a,b')| + |C(a',b) + C(a',b')| \leq 2. \qquad (1.2.11)
$$

This is (a variant of) the so-called *Bell inequality*. Given the very simple argument and the seemingly self-evident assumptions entering at various stages, one should confidently expect that it is generally obeyed by the correlations observed in any experiment of the kind described on page 4. Anything else would defy common sense, would it not? But the fact is that rather strong violations *are* observed in real-life experiments in which the left-hand side substantially exceeds 2, getting very close indeed to $2\sqrt{2}$, the maximal value allowed for Bell correlations in quantum mechanics (see Section 2.18).

Since we cannot possibly give up our convictions about locality, and thus about Einsteinian causality, the logical conclusion must be that there just is no such hidden deterministic mechanism [characterized by $\rho(\lambda)$ as well as $A_\lambda(a)$ and $B_\lambda(b)$]. We repeat:

> There is no mechanism that decides the outcome of a quantum measurement.

What is true for such correlated pairs of photons is, by inference, also true for individual photons. There is no mechanism that decides whether the photon is transmitted or reflected by the glass sheet, it is rather a *truly probabilistic* phenomenon.

This is a brutal fact of life. In a very profound sense, quantum mechanics is about learning to live with it.

1.3   Remarks on terminology

We noted the fundamental lack of determinism at the level of quantum phenomena and the consequent inability to predict the outcome of all experiments that could be performed. It may be worth emphasizing that this lack of predictive power is of a very different kind than, say, the impossibility of forecasting next year's weather.

The latter is a manifestation of the chaotic features of the underlying dynamics, frequently referred to as *deterministic chaos*. In this context, "deterministic" means that the equations of motion are differential equations that have a unique solution for given initial values — the property that we called "causal" above. There is a clash of terminology here if one wishes to diagnose one.

The deterministic chaos comes about because the solutions of the equations of motion depend extremely sensitively on the initial values, which in turn are never known with utter precision. This sensitivity is a generic feature of nonlinear equations, and not restricted to classical phenomena. Heisenberg's equations of motion of an interacting quantum system are just as nonlinear as Newton's equations for the corresponding classical system if there is one.

In classical systems that exhibit deterministic chaos, our inability to make reliable predictions concerns phenomena that are sufficiently far away in the future — a weather forecast for the next three minutes is not such a challenge. In the realm of quantum physics, however, the lack of determinism is independent of the time elapsed since the initial conditions were established. Even perfect knowledge of the state of affairs immediately before a measurement is taken does not enable us to predict the outcome; at best we can make a probabilistic prediction, a statistical prediction.

Of course, matters tend to be worse whenever one extrapolates from the present situation, which is perhaps known with satisfactory precision, to future situations. The knowledge may not be accurate enough for a long-term extrapolation. In addition to the fundamental lack of determinism in quantum physics — the nondeterministic link between the state and the phenomena — there is then a classical-type vagueness of the probabilistic predictions, rather similar to the situation of classical deterministic chaos.

Chapter 2

Kinematics:
How Quantum Systems are Described

2.1 Stern–Gerlach experiment

Now turning to the systematic development of basic concepts and, at the same time, of essential pieces of the mathematical formalism, let us consider the historical *Stern–Gerlach experiment* of 1922 (Otto Stern and Walter Gerlach, that is). In a schematic description of this experiment,

silver atoms are emerging from an oven (on the right), pass through collimating apertures, thereby forming a well defined beam of atoms, and then pass through an inhomogeneous magnetic field, eventually reaching a screen where they are collected. The inhomogeneous field is stronger at the top ($z > 0$ side) than the bottom ($z < 0$ side), which is perhaps best seen in a frontal view:

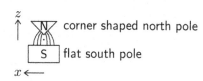

Silver atoms are endowed with a permanent magnetic dipole moment so that there is a potential energy

$$E_{\text{magn}} = -\vec{\mu} \cdot \vec{B} \tag{2.1.1}$$

associated with dipole $\vec{\mu}$ in the magnetic field $\vec{B}$. It is smallest if $\vec{\mu}$ and $\vec{B}$ are parallel, largest when they are antiparallel:

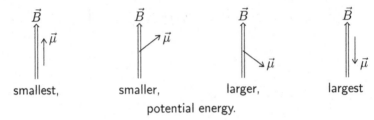

smallest,   smaller,   larger,   largest

potential energy.

As a consequence, there is a torque $\vec{\tau}$ on the dipole exerted by the magnetic field,

$$\vec{\tau} = \vec{\mu} \times \vec{B} \tag{2.1.2}$$

that tends to turn the dipole parallel to $\vec{B}$. But since the intrinsic angular momentum, the *spin*, of the atom is proportional to $\vec{\mu}$, and the rate of change of the angular momentum is just the torque, we have

$$\frac{\text{d}}{\text{d}t}\vec{\mu} \propto \vec{\mu} \times \vec{B} \tag{2.1.3}$$

which is to say that the dipole moment $\vec{\mu}$ precesses around the direction of $\vec{B}$, whereby the value of E_{magn} remains unchanged,

$$\frac{\text{d}}{\text{d}t}E_{\text{magn}} = -\frac{\text{d}\vec{\mu}}{\text{d}t} \cdot \vec{B} - \vec{\mu} \cdot \underbrace{\frac{\text{d}\vec{B}}{\text{d}t}}_{=0} \propto (\vec{\mu} \times \vec{B}) \cdot \vec{B} = 0 \,, \tag{2.1.4}$$

pictorially:

In a homogeneous magnetic field this is all that would be happening, but the field of the Stern–Gerlach magnet is inhomogeneous, its strength depends on position, mainly growing with increasing z. Therefore, the magnetic energy is position dependent,

$$E_{\mathrm{magn}}(\vec{r}) = -\vec{\mu} \cdot \vec{B}(\vec{r})\,, \qquad (2.1.5)$$

and that gives rise to a force $\vec{F}$ on the atom, given by the negative gradient of this magnetic energy,

$$\vec{F} = -\vec{\nabla} E_{\mathrm{magn}} = \vec{\nabla}\big(\vec{\mu} \cdot \vec{B}(\vec{r})\big)\,. \qquad (2.1.6)$$

For $\vec{\mu} \cdot \vec{B} > 0$, the force is toward the region of stronger $\vec{B}$ field, for $\vec{\mu} \cdot \vec{B} < 0$ it is toward the region of weaker field. Thus,

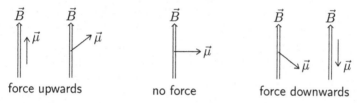

with the strength of the force proportional to the cosine of the angle between $\vec{\mu}$ and $\vec{B}$, because

$$\vec{\mu} \cdot \vec{B} = \big|\vec{\mu}\big|\big|\vec{B}\big| \cos\theta \qquad\qquad (2.1.7)$$

of course.

Now, the atoms emerging from the oven are completely unbiased in their magnetic properties, there is nothing that would prefer one orientation of $\vec{\mu}$ and discriminate against others. All orientations are equally likely which is to say that they will occur with equal frequency. Thus, some atoms will experience a strong force upwards, others a weaker one, yet others experience forces pulling downwards with a variety of strengths. Clearly, then, we expect the beam of atoms to be spread out over the screen:

But this is *not* what is observed in such an experiment, and was observed by Stern and Gerlach in 1922. Rather one finds just two spots:

It is *as if* the atoms were prealigned with the magnetic field, some having the dipole moment parallel, some having it antiparallel to $\vec{B}$. But, of course, there is no such prealignment, nothing prepares the atoms for this particular geometry of the magnetic field. For, just as well we could have chosen the dominant field component in the x direction. Then the splitting in two would be along x, perpendicular to z, and the situation would be *as if* the atoms were prealigned in the x direction. Clearly, the assumption of prophetic prealignment is ludicrous.

Rather, we have to accept that the classical prediction of a spread-out beam is inconsistent with the experimental observation. Classical physics fails here, it cannot account for the outcome of the Stern–Gerlach experiment.

Since there is nothing in the preparation of the beam that could possibly bias the atoms towards a particular direction, we expect correctly that half of the atoms are deflected up, half are deflected down. But an individual atom is not split in two, an individual atom is either deflected up or deflected down. And this happens in the perfectly probabilistic manner discussed above. There is no possibility of predicting the fate of the next atom, if we think of performing the experiment in a one-atom-at-a-time fashion.

2.2   Successive Stern–Gerlach measurements

Let us put a label on the atoms: We speak of a +atom when it is deflected up (region $z > 0$) and of a −atom when it is deflected down (region $z < 0$). In view of what we just noted, namely that we cannot possibly predict if an atom will be deflected up or down, it may not be possible to attach such labels. But in fact, there is a well defined sense to it. Consider atoms that have been deflected up by a first Stern–Gerlach magnet and are then passed through a second one:

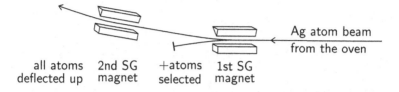

One observes (and perhaps anticipates) that all such atoms are deflected up. That is: +atoms have the objective property that they are predictably deflected up, and never down.

Likewise all −atoms are predictably deflected down:

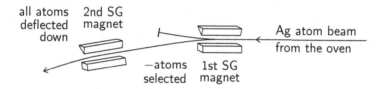

Let us simplify the drawings by a reasonable amount of abstraction:

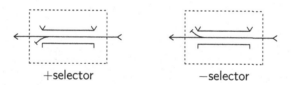

We are going to find mathematical symbols for these operations, but to do this consistently, we must first establish some basic properties.

First note that a repeated selection of the same kind is equivalent to a single selection, e.g.,

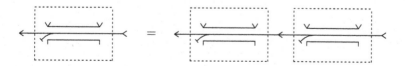

whereas a +selection followed by a −selection blocks the beam completely:

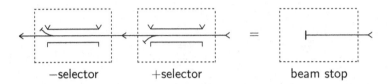

−selector   +selector   beam stop

and the same is true for a −selection followed by a +selection.

We shall also need the effect of a homogeneous magnetic field that the atom may pass through:

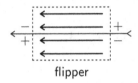

flipper

Let us take this field in the y direction and choose its strength such that the precession of the atomic magnetic moment around the y axis amounts to a rotation by 180°:

Thus, a flipper turns +atoms into −atoms, and −atoms into +atoms.

The sequence flipper, +selector, flipper is equivalent to a −selector:

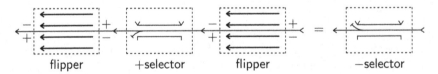

flipper   +selector   flipper   −selector

+atoms that enter the first flipper leave it as −atoms and are then rejected by the +selector, whereas −atoms that enter the first flipper leave it as +atoms and the +selector lets them pass through to the second flipper that then turns them back into −atoms. In summary, +atoms are rejected and −atoms are let through. Likewise we have

$$(\text{flipper})\,(-\text{selector})\,(\text{flipper}) = (+\text{selector})$$

when the two flippers sandwich a −selector.

2.3 Order matters

We learn something important by comparing the two orders in which we can have a single flipper and a +selector:

(a)

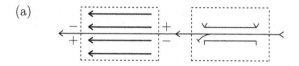

a +selector followed a by flipper accepts +atoms and turns them into −atoms;

(b)

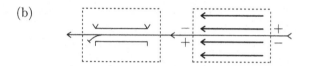

a flipper followed by a +selector accepts −atoms and turns them into +atoms.

Clearly the order matters, the two composed apparatus are very different in their net effect on the atoms.

2.4 Mathematization

This observation tells us that we need a mathematical formalism in which the apparatus (+selector, −selector, flipper, ...) are represented by symbols that can be composed (to make new apparatus), but with a composition law that is not commutative. So, clearly representing apparatus by ordinary numbers with addition or multiplication as the composition law cannot be adequate because the addition and multiplication of numbers *are* commutative — the order does *not* matter.

The simplest mathematical objects with a noncommutative composition law are square matrices and their multiplication. Presently, all we need are 2×2 matrices, for which

$$BA = \begin{pmatrix} b_{11} & b_{12} \\ b_{21} & b_{22} \end{pmatrix} \begin{pmatrix} a_{11} & a_{12} \\ a_{21} & a_{22} \end{pmatrix} = \begin{pmatrix} b_{11}a_{11} + b_{12}a_{21} & b_{11}a_{12} + b_{12}a_{22} \\ b_{21}a_{11} + b_{22}a_{21} & b_{21}a_{12} + b_{22}a_{22} \end{pmatrix}$$

$$(2.4.1)$$

and

$$AB = \begin{pmatrix} a_{11} & a_{12} \\ a_{21} & a_{22} \end{pmatrix} \begin{pmatrix} b_{11} & b_{12} \\ b_{21} & b_{22} \end{pmatrix} = \begin{pmatrix} a_{11}b_{11} + a_{12}b_{21} & a_{11}b_{12} + a_{12}b_{22} \\ a_{21}b_{11} + a_{22}b_{21} & a_{21}b_{12} + a_{22}b_{22} \end{pmatrix}$$
(2.4.2)

are different as a rule: $BA \neq AB$ (important exceptions aside).

And for the description of the state of the atom, fitting to the 2×2 matrices for the apparatus, we shall use two-component columns. The basic ones are

$$\begin{pmatrix} 1 \\ 0 \end{pmatrix} \text{ for +atoms and } \begin{pmatrix} 0 \\ 1 \end{pmatrix} \text{ for −atoms.} \tag{2.4.3}$$

We shall immediately address an important issue: What meaning shall we give to columns such as

$$\begin{pmatrix} -1 \\ 0 \end{pmatrix}, \quad \begin{pmatrix} i \\ 0 \end{pmatrix}, \quad \text{or more generally } \begin{pmatrix} \alpha \\ 0 \end{pmatrix}? \tag{2.4.4}$$

Except for the $\pm$ property — we recall that it is the objective property of being predictably deflected up or down — there is no other property of the atom to be specified. Therefore, the only plausible option is to make no difference between the columns. *All* columns of the form $\begin{pmatrix} \alpha \\ 0 \end{pmatrix}$ are equivalent. And yet we do impose one restriction: The complex number α is normalized to unit modulus, or $|\alpha|^2 = 1$. Columns such as $\begin{pmatrix} 2 \\ 0 \end{pmatrix}$ would need to be normalized, but that is a simple matter.

Likewise, all columns of the form $\begin{pmatrix} 0 \\ \beta \end{pmatrix}$ with $|\beta|^2 = 1$ are equivalent, we shall make no effort in distinguishing them from each other. Every column of this kind symbolizes a −atom.

The 2×2 matrix for the +selector is then

$$\begin{pmatrix} 1 & 0 \\ 0 & 0 \end{pmatrix} \equiv S_+ \tag{2.4.5}$$

and we verify that it does what it should: it lets +atoms pass unaffected, but rejects −atoms,

$$\begin{pmatrix} 1 & 0 \\ 0 & 0 \end{pmatrix} \begin{pmatrix} 1 \\ 0 \end{pmatrix} = \begin{pmatrix} 1 \\ 0 \end{pmatrix}, \quad \begin{pmatrix} 1 & 0 \\ 0 & 0 \end{pmatrix} \begin{pmatrix} 0 \\ 1 \end{pmatrix} = \begin{pmatrix} 0 \\ 0 \end{pmatrix} = 0. \tag{2.4.6}$$

Likewise

$$\begin{pmatrix} 0\ 0 \\ 0\ 1 \end{pmatrix} \equiv S_- \qquad (2.4.7)$$

will symbolize the $-$selector, and

$$\begin{pmatrix} 0\ 0 \\ 0\ 1 \end{pmatrix}\begin{pmatrix} 1 \\ 0 \end{pmatrix} = 0, \qquad \begin{pmatrix} 0\ 0 \\ 0\ 1 \end{pmatrix}\begin{pmatrix} 0 \\ 1 \end{pmatrix} = \begin{pmatrix} 0 \\ 1 \end{pmatrix} \qquad (2.4.8)$$

show that the assignment is consistent.

The flipper interchanges $\begin{pmatrix} 1 \\ 0 \end{pmatrix}$ and $\begin{pmatrix} 0 \\ 1 \end{pmatrix}$, so that $\begin{pmatrix} 0\ 1 \\ 1\ 0 \end{pmatrix}$ is an obvious choice. However, to be later in agreement with standard conventions, we will use

$$\begin{pmatrix} 0\ 1 \\ -1\ 0 \end{pmatrix} \equiv F \qquad (2.4.9)$$

instead, so that

$$F\begin{pmatrix} 1 \\ 0 \end{pmatrix} = \begin{pmatrix} 0 \\ -1 \end{pmatrix}, \qquad F\begin{pmatrix} 0 \\ 1 \end{pmatrix} = \begin{pmatrix} 1 \\ 0 \end{pmatrix}. \qquad (2.4.10)$$

We have already noted that there is no significant difference between $\begin{pmatrix} 0 \\ 1 \end{pmatrix}$ and $\begin{pmatrix} 0 \\ -1 \end{pmatrix}$, both stand for "$-$atom" on equal footing, and so this conventional minus sign in F is of no deeper physical significance.

We put the symbols to a couple of tests, checking if the assignments agree with what we have found earlier. First, two $+$selections in succession must be like just one,

$$S_+ S_+ = S_+ : \qquad \begin{pmatrix} 1\ 0 \\ 0\ 0 \end{pmatrix}\begin{pmatrix} 1\ 0 \\ 0\ 0 \end{pmatrix} = \begin{pmatrix} 1\ 0 \\ 0\ 0 \end{pmatrix}, \qquad (2.4.11)$$

likewise for successive $-$selections,

$$S_- S_- = S_- : \qquad \begin{pmatrix} 0\ 0 \\ 0\ 1 \end{pmatrix}\begin{pmatrix} 0\ 0 \\ 0\ 1 \end{pmatrix} = \begin{pmatrix} 0\ 0 \\ 0\ 1 \end{pmatrix}, \qquad (2.4.12)$$

and a $+$selection followed by a $-$selection must leave nothing,

$$S_- S_+ = \begin{pmatrix} 0\ 0 \\ 0\ 1 \end{pmatrix}\begin{pmatrix} 1\ 0 \\ 0\ 0 \end{pmatrix} = \begin{pmatrix} 0\ 0 \\ 0\ 0 \end{pmatrix} = 0, \qquad (2.4.13)$$

indeed. The same for the reversed order,

$$S_+S_- = \begin{pmatrix} 1 & 0 \\ 0 & 0 \end{pmatrix}\begin{pmatrix} 0 & 0 \\ 0 & 1 \end{pmatrix} = \begin{pmatrix} 0 & 0 \\ 0 & 0 \end{pmatrix} = 0. \qquad (2.4.14)$$

Note that, keeping in mind that eventually a column stands on the right, these products are such that the earlier operation stands to the right of the later one, as in

$$S_-S_+ \begin{pmatrix} 1 \\ 0 \end{pmatrix} = S_- \begin{pmatrix} 1 \\ 0 \end{pmatrix} = \begin{pmatrix} 0 \\ 0 \end{pmatrix} = 0. \qquad (2.4.15)$$

$$\downarrow \qquad\qquad\qquad\qquad \downarrow$$

$$+\text{atom entering} \qquad\qquad \text{nothing goes through}$$

Slightly more involved is the sequence

$$(\text{flipper})(-\text{selector})(\text{flipper}) = (+\text{selector}) \qquad (2.4.16)$$

where we should verify that

$$FS_+F \quad \text{has the same effect as} \quad S_-. \qquad (2.4.17)$$

Let us see:

$$FS_+F = \begin{pmatrix} 0 & 1 \\ -1 & 0 \end{pmatrix}\begin{pmatrix} 1 & 0 \\ 0 & 0 \end{pmatrix}\begin{pmatrix} 0 & 1 \\ -1 & 0 \end{pmatrix}$$

$$= \begin{pmatrix} 0 & 1 \\ -1 & 0 \end{pmatrix}\begin{pmatrix} 0 & 1 \\ 0 & 0 \end{pmatrix} = \begin{pmatrix} 0 & 0 \\ 0 & -1 \end{pmatrix} = -S_-. \qquad (2.4.18)$$

Indeed, it works, once more with an overall minus sign that must not bother us. Similarly, we should have

$$FS_-F = S_+ \qquad (2.4.19)$$

except perhaps for another minus sign of this sort. See

$$FS_-F = \begin{pmatrix} 0 & 1 \\ -1 & 0 \end{pmatrix}\begin{pmatrix} 0 & 0 \\ 0 & 1 \end{pmatrix}\begin{pmatrix} 0 & 1 \\ -1 & 0 \end{pmatrix}$$

$$= \begin{pmatrix} 0 & 1 \\ -1 & 0 \end{pmatrix}\begin{pmatrix} 0 & 0 \\ 0 & -1 \end{pmatrix} = \begin{pmatrix} -1 & 0 \\ 0 & 0 \end{pmatrix} = -S_+ \qquad (2.4.20)$$

indeed, and yes there is another minus sign appearing.

Finally, let us see how the two situations of Section 2.3 differ:

$$\text{(a) } FS_+ = \begin{pmatrix} 0 & 1 \\ -1 & 0 \end{pmatrix}\begin{pmatrix} 1 & 0 \\ 0 & 0 \end{pmatrix} = \begin{pmatrix} 0 & 0 \\ -1 & 0 \end{pmatrix},$$

$$\text{(b) } S_+F = \begin{pmatrix} 1 & 0 \\ 0 & 0 \end{pmatrix}\begin{pmatrix} 0 & 1 \\ -1 & 0 \end{pmatrix} = \begin{pmatrix} 0 & 1 \\ 0 & 0 \end{pmatrix}. \tag{2.4.21}$$

They are very different as they should be, and these situations really represent the physics correctly. See

(a) accepts +, turns it into −,

$$\begin{pmatrix} 0 & 0 \\ -1 & 0 \end{pmatrix}\begin{pmatrix} 1 \\ 0 \end{pmatrix} = \begin{pmatrix} 0 \\ -1 \end{pmatrix},$$

while rejecting −atoms,

$$\begin{pmatrix} 0 & 0 \\ -1 & 0 \end{pmatrix}\begin{pmatrix} 0 \\ 1 \end{pmatrix} = 0;$$

$$\tag{2.4.22}$$

(b) rejects +atoms,

$$\begin{pmatrix} 0 & 1 \\ 0 & 0 \end{pmatrix}\begin{pmatrix} 1 \\ 0 \end{pmatrix} = 0,$$

and accepts −atoms, turning them into +atoms,

$$\begin{pmatrix} 0 & 1 \\ 0 & 0 \end{pmatrix}\begin{pmatrix} 0 \\ 1 \end{pmatrix} = \begin{pmatrix} 1 \\ 0 \end{pmatrix}.$$

When we first introduced the "flipper" on page 16, we took the magnetic field pointing in the $+y$ direction. Just as well, we could have it pointing in the $-y$ direction, thus getting the "antiflipper,"

$$F^{-1} = \begin{pmatrix} 0 & -1 \\ 1 & 0 \end{pmatrix}. \tag{2.4.23}$$

antiflipper

Since it undoes the effect of the flipper — one turning magnetic moments clockwise, the other counter-clockwise — we use the inverse F^{-1} of the flipper matrix F to symbolize the antiflipper. One verifies easily that

$$F^{-1}F = \begin{pmatrix} 0 & -1 \\ 1 & 0 \end{pmatrix} \begin{pmatrix} 0 & 1 \\ -1 & 0 \end{pmatrix} = \begin{pmatrix} 1 & 0 \\ 0 & 1 \end{pmatrix} = 1 \,,$$

$$FF^{-1} = \begin{pmatrix} 0 & 1 \\ -1 & 0 \end{pmatrix} \begin{pmatrix} 0 & -1 \\ 1 & 0 \end{pmatrix} = \begin{pmatrix} 1 & 0 \\ 0 & 1 \end{pmatrix} = 1 \,, \qquad (2.4.24)$$

which translate into the physical statement that flipper and antiflipper compensate for each other, irrespective of the order in which they act.

2-1 Show that this commutativity is generally true. That is if X, Y, Z are $n \times n$ square matrices such that

$$XY = 1, \qquad YZ = 1$$

then it follows (how?) that $X = Z$, and we write $X = Z = Y^{-1}$.

In other words, there is a *unique inverse*, not one inverse for multiplication on the left and another for multiplication on the right.

The matrices F, F^{-1} that we found for the flipper and antiflipper should be particular cases of the matrices that stand for stretches of magnetic field, pointing in the y direction, but with varying strength. A rather weak field will not turn +atoms into −atoms, but will have some small effect on them, rotating magnetic moments by some small angle. More generally, we characterize any such field by the angle ϕ by which it rotates the magnetic moments. Then

$$R_y(\phi) = \begin{pmatrix} \cos\frac{\phi}{2} & \sin\frac{\phi}{2} \\ -\sin\frac{\phi}{2} & \cos\frac{\phi}{2} \end{pmatrix} \qquad (2.4.25)$$

is appropriate. We check the consistency of

$$R_y(\phi = 0) = 1 \,,$$
$$R_y(\phi = \pi) = F \,,$$
$$R_y(\phi = -\pi) = F^{-1} \qquad (2.4.26)$$

by inspection and verify that two successive rotations of this sort, are just one rotation by the net angle:

$$\underbrace{R_y(\phi_2)}_{\substack{\text{then rotate} \\ \text{by angle } \phi_2}} \underbrace{R_y(\phi_1)}_{\substack{\text{first rotate} \\ \text{by angle } \phi_1}} = \underbrace{R_y(\phi_1 + \phi_2)}_{\substack{\text{which is as much as} \\ \text{rotating by angle } \phi_1 + \phi_2}} \,. \qquad (2.4.27)$$

Written out, this is

$$\begin{pmatrix} \cos\frac{\phi_2}{2} & \sin\frac{\phi_2}{2} \\ -\sin\frac{\phi_2}{2} & \cos\frac{\phi_2}{2} \end{pmatrix} \begin{pmatrix} \cos\frac{\phi_1}{2} & \sin\frac{\phi_1}{2} \\ -\sin\frac{\phi_1}{2} & \cos\frac{\phi_1}{2} \end{pmatrix} = \begin{pmatrix} \cos\frac{\phi_1+\phi_2}{2} & \sin\frac{\phi_1+\phi_2}{2} \\ -\sin\frac{\phi_1+\phi_2}{2} & \cos\frac{\phi_1+\phi_2}{2} \end{pmatrix},$$

(2.4.28)

where the trigonometric addition theorems

$$\cos\frac{\phi_2}{2}\cos\frac{\phi_1}{2} - \sin\frac{\phi_2}{2}\sin\frac{\phi_1}{2} = \cos\frac{\phi_2+\phi_1}{2},$$

$$\sin\frac{\phi_2}{2}\cos\frac{\phi_1}{2} + \cos\frac{\phi_2}{2}\sin\frac{\phi_1}{2} = \sin\frac{\phi_2+\phi_1}{2}$$

(2.4.29)

are applied.

2.5 Probabilities and probability amplitudes

By looking at the effect of a rotation by $\frac{\pi}{2} = 90°$, we can now identify the columns that stand for "+atoms in the x direction" and "−atom in the x direction":

$$\underbrace{R_y\left(\frac{\pi}{2}\right)}_{\substack{\text{rotate} \\ \text{by } 90°}} \underbrace{\begin{pmatrix} 1 \\ 0 \end{pmatrix}}_{\substack{+ \text{ in the } z \\ \text{direction}}} = \frac{1}{\sqrt{2}}\begin{pmatrix} 1 & 1 \\ -1 & 1 \end{pmatrix}\begin{pmatrix} 1 \\ 0 \end{pmatrix} = \underbrace{\frac{1}{\sqrt{2}}\begin{pmatrix} 1 \\ -1 \end{pmatrix}}_{\substack{- \text{ in the } x \\ \text{direction}}},$$

$$\underbrace{R_y\left(\frac{\pi}{2}\right)}_{\substack{\text{rotate} \\ \text{by } 90°}} \underbrace{\begin{pmatrix} 0 \\ 1 \end{pmatrix}}_{\substack{- \text{ in the } z \\ \text{direction}}} = \frac{1}{\sqrt{2}}\begin{pmatrix} 1 & 1 \\ -1 & 1 \end{pmatrix}\begin{pmatrix} 1 \\ 0 \end{pmatrix} = \underbrace{\frac{1}{\sqrt{2}}\begin{pmatrix} 1 \\ 1 \end{pmatrix}}_{\substack{+ \text{ in the } x \\ \text{direction}}}. \quad (2.5.1)$$

If a Stern–Gerlach measurement is performed with the dominant component of the magnetic field in the z direction, atoms of the "± in x" kind will be deflected up or down with 50% probability each, because we get the "± in x" kind halfway between "+ in z" and "− in z":

$$\underbrace{-\begin{pmatrix} 0 \\ 1 \end{pmatrix}}_{- \text{ in } z} = R_y(\pi) \underbrace{\begin{pmatrix} 1 \\ 0 \end{pmatrix}}_{+ \text{ in } z} = \left[R_y\left(\frac{\pi}{2}\right)\right]^2 \begin{pmatrix} 1 \\ 0 \end{pmatrix}$$

$$= R_y\left(\frac{\pi}{2}\right) \underbrace{R_y\left(\frac{\pi}{2}\right)\begin{pmatrix} 1 \\ 0 \end{pmatrix}}_{\text{"halfway"}}. \quad (2.5.2)$$

Asking for the probabilities of +deflection and −deflection, we begin with the ratio $100\% : 0\%$ for $\begin{pmatrix} 0 \\ 1 \end{pmatrix}$, and have the ratio $50\% : 50\%$ halfway in between for $R_y\left(\dfrac{\pi}{2}\right)\begin{pmatrix} 1 \\ 0 \end{pmatrix} \widehat{=}$ "− in x".

Likewise

$$\begin{pmatrix} 0 \\ 1 \end{pmatrix} = R_y(-\pi)\begin{pmatrix} 1 \\ 0 \end{pmatrix} = R_y\left(-\frac{\pi}{2}\right)\underbrace{R_y\left(-\frac{\pi}{2}\right)\begin{pmatrix} 1 \\ 0 \end{pmatrix}}_{\text{"halfway"}}$$

$$= R_y\left(-\frac{\pi}{2}\right)\frac{1}{\sqrt{2}}\begin{pmatrix} 1 \\ 1 \end{pmatrix} \tag{2.5.3}$$

says the same thing about $R_y\left(-\dfrac{\pi}{2}\right)\dfrac{1}{\sqrt{2}}\begin{pmatrix} 1 \\ 0 \end{pmatrix} \widehat{=}$ "+in x". Note that

$$\text{"}\pm \text{ in } x\text{"} \widehat{=} \frac{1}{\sqrt{2}}\begin{pmatrix} 1 \\ \pm 1 \end{pmatrix} = \frac{1}{\sqrt{2}}\begin{pmatrix} 1 \\ 0 \end{pmatrix} + \frac{1}{\sqrt{2}}\begin{pmatrix} 0 \\ \pm 1 \end{pmatrix} \tag{2.5.4}$$

so that the "$\pm$ in x" states are weighted sums of the "$\pm$ in z" states, with coefficients whose *squares* are those 50% probabilities:

$$\left|\frac{1}{\sqrt{2}}\right|^2 = \frac{1}{2}. \tag{2.5.5}$$

2-2 Compare the following situations.

(1) A beam of atoms, half of which are preselected as "+ in z", the other half as "− in z", is sent through a Stern–Gerlach apparatus that
   (i) sorts in the z direction,
   (ii) sorts in the x direction.
(2) A beam of atoms, all of which are preselected as "+ in x", is sent through a Stern–Gerlach apparatus that
   (i) sorts in the z direction,
   (ii) sorts in the x direction.

Do the measurement results enable you to tell apart situations 1 and 2?

More generally, we have as the result of a rotation by ϕ

$$R_y(\phi)\begin{pmatrix} 1 \\ 0 \end{pmatrix} = \begin{pmatrix} \cos\frac{\phi}{2} \\ -\sin\frac{\phi}{2} \end{pmatrix} = \cos\frac{\phi}{2}\begin{pmatrix} 1 \\ 0 \end{pmatrix} + \sin\frac{\phi}{2}\begin{pmatrix} 0 \\ -1 \end{pmatrix} \tag{2.5.6}$$

so that $\begin{pmatrix} 1 \\ 0 \end{pmatrix}$ and $\begin{pmatrix} 0 \\ -1 \end{pmatrix}$ are summed with weights $\cos\frac{\phi}{2}$ and $\sin\frac{\phi}{2}$, whose *squares* add up to unit,

$$\left(\cos\frac{\phi}{2}\right)^2 + \left(\sin\frac{\phi}{2}\right)^2 = 1. \tag{2.5.7}$$

This invites the following interpretation:

The fraction $(\cos\frac{\phi}{2})^2 = \frac{1}{2}(1+\cos\phi)$ of these atoms will be deflected in the $+z$ direction, and the fraction $(\sin\frac{\phi}{2})^2 = \frac{1}{2}(1-\cos\phi)$ will be deflected downwards.

We can regard the statements as *probabilistic predictions* of our still under-developed formalism. We should set out and verify them by the appropriate experiment, namely,

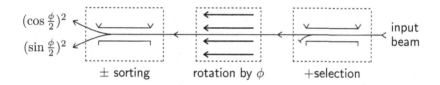

and such an experiment would confirm the predictions.

Having thus identified columns for "$\pm$ in z" and "$\pm$ in x", how about "$\pm$ in y"? Atoms that have their magnetic moments aligned with the y axis, be it parallel or antiparallel, are not rotated by the flipper at all, so that

$$F\begin{pmatrix} \alpha \\ \beta \end{pmatrix} = \lambda\begin{pmatrix} \alpha \\ \beta \end{pmatrix} \quad \text{if} \begin{pmatrix} \alpha \\ \beta \end{pmatrix} \text{ is "+ in } y\text{" or "− in } y\text{"} \tag{2.5.8}$$

where λ is any complex number. With $F = \begin{pmatrix} 0 & 1 \\ -1 & 0 \end{pmatrix}$ this says

$$\left.\begin{array}{r} \beta = \lambda\alpha \\ -\alpha = \lambda\beta \end{array}\right\} \quad \text{or} \quad -\alpha\beta = \lambda^2\alpha\beta. \tag{2.5.9}$$

The solutions $\alpha = 0$ or $\beta = 0$ are not permitted because $\begin{pmatrix} 0 \\ \beta \end{pmatrix}$ and $\begin{pmatrix} \alpha \\ 0 \end{pmatrix}$ are "− in z" and "+ in z" respectively, so we must have

$$\lambda = +\mathrm{i} \quad \text{or} \quad \lambda = -\mathrm{i}. \tag{2.5.10}$$

In the first case we get

$$\frac{1}{\sqrt{2}}\begin{pmatrix} 1 \\ i \end{pmatrix} \cong \text{``+ in } y\text{''}, \tag{2.5.11}$$

and

$$\frac{1}{\sqrt{2}}\begin{pmatrix} 1 \\ -i \end{pmatrix} \cong \text{``- in } y\text{''} \tag{2.5.12}$$

results in the second case. The factors of $\frac{1}{\sqrt{2}}$ are supplied in analogy with the x states, so that *squaring* the prefactors in

$$\frac{1}{\sqrt{2}}\begin{pmatrix} 1 \\ \pm i \end{pmatrix} = \frac{1}{\sqrt{2}}\begin{pmatrix} 1 \\ 0 \end{pmatrix} \pm \frac{i}{\sqrt{2}}\begin{pmatrix} 0 \\ 1 \end{pmatrix} \tag{2.5.13}$$

again gives the probabilities of

$$\left|\frac{1}{\sqrt{2}}\right|^2 = \frac{1}{2} \quad \text{and} \quad \left|\pm\frac{i}{\sqrt{2}}\right|^2 = \frac{1}{2} \tag{2.5.14}$$

for deflection up or down in a z-deflection Stern–Gerlach measurement.

Consistency requires that all spatial directions are on equal footing. So, atoms prepared (= preselected) as "– in y" or "+ in y" should also be deflected 50% : 50% in a x deflection measurement. According to the rules just established, we verify this by writing the "$\pm$ in y" states as weighted sums of the "$\pm$ in x" states and then *square* the coefficients. Let us see:

$$\underbrace{\frac{1}{\sqrt{2}}\begin{pmatrix} 1 \\ \pm i \end{pmatrix}}_{\pm \text{ in } y} = \underbrace{\frac{1\pm i}{2}\frac{1}{\sqrt{2}}\begin{pmatrix} 1 \\ 1 \end{pmatrix}}_{+ \text{ in } x} + \underbrace{\frac{1\mp i}{2}\frac{1}{\sqrt{2}}\begin{pmatrix} 1 \\ -1 \end{pmatrix}}_{- \text{ in } x}, \tag{2.5.15}$$

so the probabilities in question are

$$\left|\frac{1\pm i}{2}\right|^2 = \frac{(1\mp i)(1\pm i)}{4} = \frac{1}{2}, \quad \text{indeed.} \tag{2.5.16}$$

2-3 Show that one also gets 50% : 50% probability when atoms prese-lected as "+ in x" or "– in x" are sent through a y-sorting Stern–Gerlach apparatus.

2-4 How would you use a homogeneous magnetic field in the x direction combined with a z-deflecting Stern–Gerlach magnet to carry out the $\pm y$ sorting?

Consistent with (2.5.4) and (2.5.5) as well as (2.5.13) and (2.5.14), and more general than (2.5.6) and (2.5.7), we normalize an arbitrary column $\begin{pmatrix} \alpha \\ \beta \end{pmatrix}$ by

$$|\alpha|^2 + |\beta|^2 = 1, \qquad (2.5.17)$$

with the ultimate justification given below when we arrive at (2.5.38). How do we accomplish the presentation of such a column in terms of the "$\pm$ in x", or "$\pm$ in y", or "$\pm$ in z" columns quite generally? It is easy for z:

$$\begin{pmatrix} \alpha \\ \beta \end{pmatrix} = \alpha \begin{pmatrix} 1 \\ 0 \end{pmatrix} + \beta \begin{pmatrix} 0 \\ 1 \end{pmatrix} \qquad (2.5.18)$$

with coefficients

$$\alpha = (1, 0) \begin{pmatrix} \alpha \\ \beta \end{pmatrix} \qquad (2.5.19)$$

and

$$\beta = (0, 1) \begin{pmatrix} \alpha \\ \beta \end{pmatrix}. \qquad (2.5.20)$$

For the x columns we have

$$\begin{pmatrix} \alpha \\ \beta \end{pmatrix} = \frac{\alpha + \beta}{\sqrt{2}} \frac{1}{\sqrt{2}} \begin{pmatrix} 1 \\ 1 \end{pmatrix} + \frac{\alpha - \beta}{\sqrt{2}} \frac{1}{\sqrt{2}} \begin{pmatrix} 1 \\ -1 \end{pmatrix} \qquad (2.5.21)$$

with coefficients

$$\frac{\alpha + \beta}{\sqrt{2}} = \frac{1}{\sqrt{2}} (1, 1) \begin{pmatrix} \alpha \\ \beta \end{pmatrix} \qquad (2.5.22)$$

and

$$\frac{\alpha - \beta}{\sqrt{2}} = \frac{1}{\sqrt{2}} (1, -1) \begin{pmatrix} \alpha \\ \beta \end{pmatrix}. \qquad (2.5.23)$$

Finally, for the "$\pm$ in y" columns we have

$$\begin{pmatrix} \alpha \\ \beta \end{pmatrix} = \frac{\alpha - i\beta}{\sqrt{2}} \frac{1}{\sqrt{2}} \begin{pmatrix} 1 \\ i \end{pmatrix} + \frac{\alpha + i\beta}{\sqrt{2}} \frac{1}{\sqrt{2}} \begin{pmatrix} 1 \\ -i \end{pmatrix} \qquad (2.5.24)$$

with coefficients

$$\frac{\alpha - i\beta}{\sqrt{2}} = \frac{1}{\sqrt{2}} (1, -i) \begin{pmatrix} \alpha \\ \beta \end{pmatrix} \qquad (2.5.25)$$

and

$$\frac{\alpha + i\beta}{\sqrt{2}} = \frac{1}{\sqrt{2}} (1, i) \binom{\alpha}{\beta}. \tag{2.5.26}$$

Clearly, there is a general pattern here, namely the coefficients of one particular decomposition are obtained by multiplying $\binom{\alpha}{\beta}$ with the corresponding *rows*, obtained by taking the adjoints of the *columns* in questions, that is: the complex conjugate of the transposed columns. It is important here that the columns we use for the weighted sum are orthogonal to each other. In general, then, we have this. There are two columns

$$\binom{a_1}{b_1} \quad \text{and} \quad \binom{a_2}{b_2} \tag{2.5.27}$$

that are normalized

$$|a_1|^2 + |b_1|^2 = 1, \qquad |a_2|^2 + |b_2|^2 = 1 \tag{2.5.28}$$

and orthogonal to each other,

$$0 = \binom{a_1}{b_1}^\dagger \binom{a_2}{b_2} = (a_1^*, b_1^*) \binom{a_2}{b_2} = a_1^* a_2 + b_1^* b_2 \tag{2.5.29}$$

or, equivalently after complex conjugation,

$$0 = \binom{a_2}{b_2}^\dagger \binom{a_1}{b_1} = (a_2^*, b_2^*) \binom{a_1}{b_1} = a_2^* a_1 + b_2^* b_1, \tag{2.5.30}$$

where the dagger † indicates the adjoint. Then

$$\binom{\alpha}{\beta} = \binom{a_1}{b_1} (a_1^*, b_1^*) \binom{\alpha}{\beta} + \binom{a_2}{b_2} (a_2^*, b_2^*) \binom{\alpha}{\beta} \tag{2.5.31}$$

as we can verify by multiplying from the left by $(a_1^*, b_1^*) = \binom{a_1}{b_1}^\dagger$ and $(a_2^*, b_2^*) = \binom{a_2}{b_2}^\dagger$, respectively. All particular decompositions in (2.5.18)–(2.5.26) are examples for this general rule, for the pairs of columns that stand for "± in z", "± in x", and "± in y".

But there is yet another way of looking at this. So far we have read

$$\underbrace{\begin{pmatrix} a_1 \\ b_1 \end{pmatrix}}_{\substack{\text{column in the} \\ \text{decomposition}}} \underbrace{(a_1^*, b_1^*) \begin{pmatrix} \alpha \\ \beta \end{pmatrix}}_{\substack{\text{coefficient that goes} \\ \text{with this column}}} . \tag{2.5.32}$$

Now let us choose another association, which is allowed because matrix multiplication is associative,

$$\underbrace{\begin{pmatrix} a_1 \\ b_1 \end{pmatrix} (a_1^*, b_1^*)}_{\substack{\text{matrix that projects} \\ \text{on column } \begin{pmatrix} a_1 \\ b_1 \end{pmatrix}}} \underbrace{\begin{pmatrix} \alpha \\ \beta \end{pmatrix}}_{\text{input}} . \tag{2.5.33}$$

The *projection* property is verified by

$$\left[\begin{pmatrix} a_1 \\ b_1 \end{pmatrix} (a_1^*, b_1^*) \right] \begin{pmatrix} a_1 \\ b_1 \end{pmatrix} = \begin{pmatrix} a_1 \\ b_1 \end{pmatrix} \underbrace{\left[(a_1^*, b_1^*) \begin{pmatrix} a_1 \\ b_1 \end{pmatrix} \right]}_{= 1 \text{ by normalization}} = \begin{pmatrix} a_1 \\ b_1 \end{pmatrix} \tag{2.5.34}$$

and

$$\left[\begin{pmatrix} a_1 \\ b_1 \end{pmatrix} (a_1^*, b_1^*) \right] \begin{pmatrix} a_2 \\ b_2 \end{pmatrix} = \begin{pmatrix} a_1 \\ b_1 \end{pmatrix} \underbrace{\left[(a_1^*, b_1^*) \begin{pmatrix} a_2 \\ b_2 \end{pmatrix} \right]}_{= 0 \text{ by orthogonality}} = 0 , \tag{2.5.35}$$

indeed.

And, finally, upon reading the decomposition (2.5.31) as involving the sum of two projectors,

$$\begin{pmatrix} \alpha \\ \beta \end{pmatrix} = \left[\underbrace{\begin{pmatrix} a_1 \\ b_1 \end{pmatrix} (a_1^*, b_1^*)}_{\substack{\text{projects on} \\ \text{column } \begin{pmatrix} a_1 \\ b_1 \end{pmatrix}}} + \underbrace{\begin{pmatrix} a_1 \\ b_1 \end{pmatrix} (a_1^*, b_1^*)}_{\substack{\text{projects on} \\ \text{column } \begin{pmatrix} a_1 \\ b_1 \end{pmatrix}}} \right] \begin{pmatrix} \alpha \\ \beta \end{pmatrix} , \tag{2.5.36}$$

we note that the sum of the two projection matrices must be the unit matrix,

$$\begin{pmatrix} a_1 \\ b_1 \end{pmatrix} (a_1^*, b_1^*) + \begin{pmatrix} a_2 \\ b_2 \end{pmatrix} (a_2^*, b_2^*) = \begin{pmatrix} 1 & 0 \\ 0 & 1 \end{pmatrix} = 1 . \tag{2.5.37}$$

2-5 Illustrate this by writing out explicitly the six projection matrices for the "$\pm$ in x", "$\pm$ in y", and "$\pm$ in z" columns, and adding each pair.

2-6 Exploit the normalization and orthogonality conditions of (2.5.28)–(2.5.30) to show that the statement is generally correct. Hint: Consider the 2×2 matrix $U = \begin{pmatrix} a_1 & a_2 \\ b_1 & b_2 \end{pmatrix}$ and show first that $U^\dagger U = 1$, then take a look at $UU^\dagger$.

Harkening back to pages 24–26, we recall that in all particular examples discussed then, *probabilities* for deflection in the respective $+$ and $-$ directions were always correctly given by the absolute squares of the coefficients. This is why the coefficients are termed *probability amplitudes*. They are explicitly stated in (2.5.18)–(2.5.26), and we find the resulting probabilities as

$$\text{prob}(+ \text{ in } z) = |\alpha|^2, \qquad \text{prob}(- \text{ in } z) = |\beta|^2,$$

$$\text{prob}(\pm \text{ in } x) = \left| \frac{\alpha \pm \beta}{\sqrt{2}} \right|^2 = \frac{1}{2}(|\alpha|^2 + |\beta|^2 \pm \alpha^*\beta \pm \beta^*\alpha)$$
$$= \frac{1}{2} \pm \text{Re}(\alpha^*\beta),$$

$$\text{prob}(\pm \text{ in } y) = \left| \frac{\alpha \mp i\beta}{\sqrt{2}} \right|^2 = \frac{1}{2}(|\alpha|^2 + |\beta|^2 \mp i\alpha^*\beta \pm i\beta^*\alpha)$$
$$= \frac{1}{2} \pm \text{Im}(\alpha^*\beta), \qquad (2.5.38)$$

where the normalization $|\alpha|^2 + |\beta|^2 = 1$ of (2.5.17) enters, which we now recognize as the obvious condition that the probabilities for "$+$ in z" and "$-$ in z" must add up to 100%.

The operational meaning of these statements is as follows. Suppose we have atoms $+$ selected in the z direction and then passed through a homogeneous magnetic field of unknown strength and direction,

so that the atoms that emerge are described by some column $\begin{pmatrix} \alpha \\ \beta \end{pmatrix}$, but we do not know the values of α and β. How can we determine them? By performing a z-deflection Stern–Gerlach experiment on a fraction of them (say $\frac{1}{3}$ of all the atoms), an x-Stern–Gerlach measurement on another fraction (another $\frac{1}{3}$, say) and a y-Stern–Gerlach measurement on the rest:

$$
\begin{array}{l}
|\alpha|^2 \; + \; \boxed{z\ \mathrm{SG}} \\
|\beta|^2 \; - \\
\end{array}
\qquad 1/3
$$

$$
\frac{1}{2} \pm \mathrm{Re}(\alpha^*\beta) \;\; \begin{array}{l} + \\ - \end{array} \boxed{x\ \mathrm{SG}} \quad 1/3 \qquad\qquad \binom{\alpha}{\beta}
$$

$$
\frac{1}{2} \pm \mathrm{Im}(\alpha^*\beta) \;\; \begin{array}{l} + \\ - \end{array} \boxed{y\ \mathrm{SG}} \qquad 1/3
$$

Since $|\alpha|^2 + |\beta|^2 = 1$, the z measurement tells us the value of $|\alpha|^2 - |\beta|^2$. It is a number between -1 and $+1$, and we parameterize it conveniently by writing

$$
|\alpha|^2 - |\beta|^2 = \cos(2\theta), \qquad 0 \le \theta \le \frac{\pi}{2}. \tag{2.5.39}
$$

Then

$$
\begin{aligned}
|\alpha|^2 &= \frac{1}{2}(|\alpha|^2 + |\beta|^2) + \frac{1}{2}(|\alpha|^2 - |\beta|^2) \\
&= \frac{1}{2} + \frac{1}{2}\cos(2\theta) = (\cos\theta)^2
\end{aligned} \tag{2.5.40}
$$

and

$$
\begin{aligned}
|\beta|^2 &= \frac{1}{2}(|\alpha|^2 + |\beta|^2) - \frac{1}{2}(|\alpha|^2 - |\beta|^2) \\
&= \frac{1}{2} - \frac{1}{2}\cos(2\theta) = (\sin\theta)^2.
\end{aligned} \tag{2.5.41}
$$

Consequently, we have

$$
\alpha = e^{i\varphi}\cos\theta, \qquad \beta = e^{i\phi}\sin\theta, \tag{2.5.42}
$$

where $e^{i\varphi}$, $e^{i\phi}$ are as yet undetermined phase factors, that is complex numbers of unit modulus,

$$
\left| e^{i\varphi} \right| = |\cos\varphi + i\sin\varphi| = \sqrt{(\cos\varphi)^2 + (\sin\varphi)^2} = 1. \tag{2.5.43}
$$

We have here the first of many applications of Euler's identity

$$e^{i\varphi} = \cos\varphi + i\sin\varphi, \tag{2.5.44}$$

one of the numerous discoveries by Leonhard Euler.

Then taking the values of $\mathrm{Re}(\alpha^*\beta)$ and $\mathrm{Im}(\alpha^*\beta)$ from the x and y measurements, we establish the value of

$$\alpha^*\beta = \mathrm{Re}(\alpha^*\beta) + i\,\mathrm{Im}(\alpha^*\beta) \tag{2.5.45}$$

and compare with

$$\begin{aligned}
\alpha^*\beta &= (e^{i\varphi}\cos\theta)^*(e^{i\phi}\sin\theta) = e^{i(\phi-\varphi)}\cos\theta\sin\theta \\
&= \frac{1}{2}e^{i(\phi-\varphi)}\sin(2\theta).
\end{aligned} \tag{2.5.46}$$

Since we have found $\cos(2\theta)$ and thus $\sin(2\theta) \geq 0$ earlier from the data of the z measurement, we can now determine the value of $e^{i(\phi-\varphi)}$, and then know

$$\begin{pmatrix} \alpha \\ \beta \end{pmatrix} = e^{i\varphi}\begin{pmatrix} \cos\theta \\ e^{i(\phi-\varphi)}\sin\theta \end{pmatrix} \tag{2.5.47}$$

except for the prefactor of $e^{i\varphi}$. But this prefactor is an overall phase factor of no physical significance, so (1) we do not need it and (2) it is consistent with this lack of significance that it cannot be determined from experimental data. Indeed, we can adopt any convenient convention, such as putting $e^{i\varphi} = 1$ or $e^{i\varphi} = e^{-i(\phi-\varphi)}$, all choices being equally good.

2.6   Quantum Zeno effect

Here is a little application of what we have learned, to a somewhat amusing, but also instructive situation. Imagine atoms selected as "+ in z", that is $\begin{pmatrix} 1 \\ 0 \end{pmatrix}$, and then propagating through a magnetic field that gradually turns them into $\begin{pmatrix} 0 \\ -1 \end{pmatrix}$ with $R_y(\vartheta)\begin{pmatrix} 1 \\ 0 \end{pmatrix} = \begin{pmatrix} \cos\frac{\vartheta}{2} \\ -\sin\frac{\vartheta}{2} \end{pmatrix}$ at intermediate stages, eventually followed by a final z measurement:

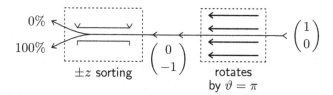

All atoms will be deflected as "$-$ in z" at the end, none as "$+$ in z". But now let us introduce a control measurement that selects "$+$ in z" halfway along the way:

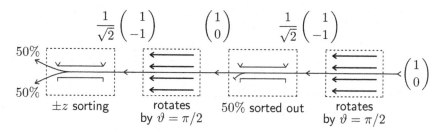

The probability that an entering atom is eventually deflected as "$+$ in z" is $50\% \times 50\% = 25\%$, because it has a 50% chance of surviving the midway sorting and another 50% of up-deflection at the end.

As a generalization, let us now break up the magnetic field in n portions, with "$-$ in z" selectors in between:

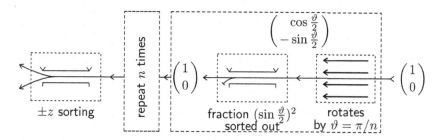

After passing through a single stretch of the field we have

$$R_y\left(\frac{\pi}{n}\right)\begin{pmatrix} 1 \\ 0 \end{pmatrix} = \begin{pmatrix} \cos\frac{\pi}{2n} \\ -\sin\frac{\pi}{2n} \end{pmatrix} \qquad (2.6.1)$$

so that the probability of making it to the next stage is

$$\left| (1,0) \begin{pmatrix} \cos\frac{\pi}{2n} \\ -\sin\frac{\pi}{2n} \end{pmatrix} \right|^2 = \left(\cos\frac{\pi}{2n} \right)^2 . \tag{2.6.2}$$

Since there are n stages of this sort all together, we have

$$p_n = \left[\left(\cos\frac{\pi}{2n} \right)^2 \right]^n = \left(\cos\frac{\pi}{2n} \right)^{2n} \tag{2.6.3}$$

for the total probability of getting through.

2-7   Calculate this probability for $n = 4, 8, 16$, and 32.

You will see that starting from $p_1 = 0$, $p_2 = \frac{1}{4}$, it increases rapidly, and approaches unity for $n \to \infty$. Of course, this limit itself is unphysical, but large values of n are not, so let us see what happens for $n \gg 1$, when $\phi = \frac{\pi}{2n}$ is very small and $\cos\phi \cong 1 - \frac{1}{2}\phi^2$ is a permissible approximation:

$$\left(\cos\frac{\pi}{2n} \right)^{2n} \cong \left[1 - \frac{1}{2}\left(\frac{\pi}{2n} \right)^2 \right]^{2n}$$

$$\cong \left[e^{-\frac{1}{2}(\frac{\pi}{2n})^2} \right]^{2n} = e^{-\frac{\pi^2}{4n}}$$

$$\cong 1 - \frac{\pi^2}{4n} . \tag{2.6.4}$$

This says that the survival probability is very close to 100% if we only take n sufficiently large.

2-8   Compare this approximation for p_n with the exact numbers for $n = 2, 4, 8, 16$, and 32.

Physically speaking, there is a magnetic field that all by itself would turn "+ in z" $\widehat{=} \begin{pmatrix} 1 \\ 0 \end{pmatrix}$ into "− in z" $\widehat{=} \begin{pmatrix} 0 \\ -1 \end{pmatrix}$, but the repeated checking "is it still $\begin{pmatrix} 1 \\ 0 \end{pmatrix}$ or already $\begin{pmatrix} 0 \\ -1 \end{pmatrix}$?" effectively "freezes" the evolution, and for $n \gg 1$ it is as if there were no magnetic field at all. This freezing of the motion is reminiscent of one of the four famous paradoxes invented by Zeno of Elea (about 450 BC), and the quantum version just discussed has become known as the *quantum Zeno effect*. It is a real physical phenomenon that is easy to demonstrate if one uses polarized photons rather than magnetic atoms.

But, putting this folklore aside, what we learn from this discussion is that a measurement disturbs the system. The evolution of $\begin{pmatrix} 1 \\ 0 \end{pmatrix}$ to $\begin{pmatrix} 0 \\ -1 \end{pmatrix}$ does not proceed as usual if you do not let the atoms alone. Every interference with their evolution disturbs or even disrupts it. You cannot check some properties of an atomic system without introducing a sizeable disturbance. This is quite different from the situation in classical physics where you can measure, say, the temperature of the water in a glass, without altering the water temperature (or any other property) noticeably.

2-9 Atoms that have been preselected as "+ in z" are successively passed through first a "+ in x" selector, then a "− in z" selector. Which fraction of the atoms is let through?

2.7 Kets and bras

Let us return to the statements on page 27,

$$
\begin{aligned}
\begin{pmatrix} \alpha \\ \beta \end{pmatrix} &= \alpha \begin{pmatrix} 1 \\ 0 \end{pmatrix} + \beta \begin{pmatrix} 0 \\ 1 \end{pmatrix} && (z \text{ columns}) \\
&= \frac{\alpha + \beta}{\sqrt{2}} \frac{1}{\sqrt{2}} \begin{pmatrix} 1 \\ 1 \end{pmatrix} + \frac{\alpha - \beta}{\sqrt{2}} \frac{1}{\sqrt{2}} \begin{pmatrix} 1 \\ -1 \end{pmatrix} && (x \text{ columns}) \\
&= \frac{\alpha - \mathrm{i}\beta}{\sqrt{2}} \frac{1}{\sqrt{2}} \begin{pmatrix} 1 \\ \mathrm{i} \end{pmatrix} + \frac{\alpha + \mathrm{i}\beta}{\sqrt{2}} \frac{1}{\sqrt{2}} \begin{pmatrix} 1 \\ -\mathrm{i} \end{pmatrix} && (y \text{ columns}) \quad (2.7.1)
\end{aligned}
$$

which decompose the general column $\begin{pmatrix} \alpha \\ \beta \end{pmatrix}$ in three different ways that refer to "± in z", "± in x", and "± in y", respectively. This has a familiar ring; it reminds us, and quite correctly so, of the representation of one and the same three-dimensional vector by numerical coefficients (sets of three) with respect to a prechosen coordinate system,

$$
\begin{aligned}
\vec{r} &= x\vec{e}_1 + y\vec{e}_2 + z\vec{e}_3 \\
&= x'\vec{e}_1' + y'\vec{e}_2' + z'\vec{e}_3'.
\end{aligned} \quad (2.7.2)
$$

It is one and the same vector $\vec{r}$, but its numerical representation,

$$\vec{r} \cong \begin{pmatrix} x \\ y \\ z \end{pmatrix} \quad \text{or} \quad \vec{r} \cong \begin{pmatrix} x' \\ y' \\ z' \end{pmatrix}, \tag{2.7.3}$$

may involve rather different sets of three numbers, depending on which co-ordinate system, which set of unit vectors they refer to. Thus it is essential that we do not confuse the vector $\vec{r}$ with its coordinates. The vector is a geometrical object all by itself, whereas its coordinates also contain reference to the coordinate system, and that we can choose quite arbitrarily, doing whatever is convenient for the calculation.

Matters are quite analogous for the magnetic atoms. The columns $\begin{pmatrix} \alpha \\ \beta \end{pmatrix}$ that we have been using make explicit reference to Stern–Gerlach measurements in what we call the z direction. But that is, of course, just as arbitrary as any such convention. Equally well we could have singled out the x direction or any other one. Thus collecting the pairs of probability amplitudes in the three decompositions in (2.7.1), we can say that atoms prepared in a particular manner (as, for example, indicated by the picture on page 30) are equivalently characterized by the columns

$$\begin{pmatrix} \alpha \\ \beta \end{pmatrix}_z, \quad \frac{1}{\sqrt{2}} \begin{pmatrix} \alpha + \beta \\ \alpha - \beta \end{pmatrix}_x, \quad \frac{1}{\sqrt{2}} \begin{pmatrix} \alpha - i\beta \\ \alpha + i\beta \end{pmatrix}_y \tag{2.7.4}$$

where the subscript indicates the reference columns that are weighted in (2.7.1).

To free ourselves from the eternal reference to particular basic columns, which are always subject to conventions and never free of arbitrariness, we take a step that is analogous to going from the numerical component column $\begin{pmatrix} x \\ y \\ z \end{pmatrix}$ to vector $\vec{r}$. The *state vectors* in the quantum-mechanical formalism are denoted by $|\ \rangle$, with appropriate identifying labels inside, and are called *kets* or ket vectors. As a reminder of the arrows we draw for magnetic moments, we use $\uparrow_z$, $\downarrow_z$ etc. as the identifying labels. Thus

$$\text{atom "+ in } z\text{": } |\uparrow_z\rangle, \quad \text{atom "− in } z\text{": } |\downarrow_z\rangle, \tag{2.7.5}$$

and likewise $|\uparrow_x\rangle$, $|\downarrow_x\rangle$ for "+ in x", "− in x" as well as $|\uparrow_y\rangle$, $|\downarrow_y\rangle$ for atoms that are "± in y".

The three decompositions of a general column in (2.7.1) are now presented as

$$| \, \rangle = |\uparrow_z\rangle \alpha + |\downarrow_z\rangle \beta = |\uparrow_x\rangle \frac{\alpha + \beta}{\sqrt{2}} + |\downarrow_x\rangle \frac{\alpha - \beta}{\sqrt{2}}$$

$$= |\uparrow_y\rangle \frac{\alpha - i\beta}{\sqrt{2}} + |\downarrow_y\rangle \frac{\alpha + i\beta}{\sqrt{2}} . \tag{2.7.6}$$

We are here expressing *one and the same ket* $| \, \rangle$ in terms of different reference kets, fully analogous to $\vec{r} = \cdots$ in (2.7.2).

Since the abstract kets are numerically represented by columns as we used them so far, they inherit in full their linear vector properties, that is we can multiply kets with complex numbers to make other kets, and we can form weighted sums of them and get more kets in this manner. The weights that appear in the sums in (2.7.6) continue to have, of course, the physical meaning that we have identified before: They are the probability amplitudes that characterize $| \, \rangle$ and tell us, upon squaring, the probabilities of $\uparrow_z$ and $\downarrow_z$, or $\uparrow_x$ and $\downarrow_x$, or $\uparrow_y$ and $\downarrow_y$.

2-10 Determine these probability amplitudes, and the resulting probabilities, for $\alpha = \frac{3}{5}$, $\beta = \frac{4}{5}$.

The systematic method for calculating probability amplitudes for given columns $\begin{pmatrix} \alpha \\ \beta \end{pmatrix}$ is, recall the discussion on page 27, by multiplying with appropriate rows from the left. These rows are obtained as hermitian conjugates, or adjoints, of the respective columns. We thus introduce yet another sort of vectors, the adjoints of the kets, denoted by $\langle \, |$ and called *bras*,

$$| \, \rangle^\dagger = \langle \, |, \qquad \langle \, |^\dagger = | \, \rangle, \tag{2.7.7}$$

and, as the second equation states, the kets are the adjoints of the bras. It follows, perhaps not unexpectedly, that taking the adjoint twice amounts to doing nothing.

Taking the adjoint is essentially a linear operation, but we must remember about the complex conjugation that is involved, so that we have

$$\left(|1\rangle a_1 + |2\rangle a_2 + |3\rangle a_3 \right)^\dagger = a_1^* \langle 1| + a_2^* \langle 2| + a_3^* \langle 3| \tag{2.7.8}$$

for the adjoint of a weighted sum of three kets, for example. More specifically, the adjoint statement to (2.7.6) is

$$\langle \, | = | \, \rangle^\dagger = \alpha^* \langle \uparrow_z| + \beta^* \langle \downarrow_z|$$
$$= \frac{\alpha^* + \beta^*}{\sqrt{2}} \langle \uparrow_x| + \frac{\alpha^* - \beta^*}{\sqrt{2}} \langle \downarrow_x|$$
$$= \frac{\alpha^* + i\beta^*}{\sqrt{2}} \langle \uparrow_y| + \frac{\alpha^* - i\beta^*}{\sqrt{2}} \langle \downarrow_y| \,. \qquad (2.7.9)$$

It should be clear that bras and kets are on equal footing. Whatever physical fact can be phrased as a relation among kets can just as well be formulated as the adjoint relation among bras.

2.8 Brackets, bra-kets, and ket-bras

Looking at, for example, the second line in (2.7.6), and its numerical version in (2.7.1), we note that the probability amplitudes for $\uparrow_x$, $\downarrow_x$ are

$$\frac{\alpha \pm \beta}{\sqrt{2}} = \frac{1}{\sqrt{2}}(1, \pm 1)\begin{pmatrix} \alpha \\ \beta \end{pmatrix} \qquad (2.8.1)$$

which translates into

$$\frac{\alpha + \beta}{\sqrt{2}} = \langle \uparrow_x| \, | \, \rangle \equiv \langle \uparrow_x| \, \rangle \,,$$
$$\frac{\alpha - \beta}{\sqrt{2}} = \langle \downarrow_x| \, | \, \rangle \equiv \langle \downarrow_x| \, \rangle \,, \qquad (2.8.2)$$

where we have adopted the usual convention to not write two vertical lines where bra meets ket: $\langle 1|\text{times}|2\rangle = \langle 1|2\rangle$, corresponding to the numerical row-times-column product.

It follows that the normalization of the basic kets to "unit length" is expressed by

$$\langle \uparrow_z | \uparrow_z \rangle = 1 \,, \qquad \langle \downarrow_z | \downarrow_z \rangle = 1 \qquad (2.8.3)$$

and their orthogonality by

$$\langle \uparrow_z | \downarrow_z \rangle = 0 \,, \qquad \langle \downarrow_z | \uparrow_z \rangle = 0 \,, \qquad (2.8.4)$$

and analogously for the x and y pairs of kets and bras. If you wonder about the origin of the somewhat frivolous terminology, introduced by Paul A. M.

Dirac, the observation

$$\underbrace{\langle 1|}_{\text{bra}} \times \underbrace{|2\rangle}_{\text{times}\ \text{ket}} = \underbrace{\langle 1|\ |2\rangle}_{\text{bra-ket}} = \underbrace{\langle 1|2\rangle}_{\text{bracket}} \qquad (2.8.5)$$

should tell you.

So, we have

$$|\ \rangle = |\uparrow_z\rangle\alpha + |\downarrow_z\rangle\beta = \left(|\uparrow_z\rangle, |\downarrow_z\rangle\right)\begin{pmatrix} \alpha \\ \beta \end{pmatrix} \qquad (2.8.6)$$

where we exhibit the reference kets as a two-component row of kets and the probability amplitudes as a numerical two-component column. Then

$$\begin{pmatrix} \alpha \\ \beta \end{pmatrix} = \begin{pmatrix} \langle\uparrow_z|\ \rangle \\ \langle\downarrow_z|\ \rangle \end{pmatrix} = \begin{pmatrix} \langle\uparrow_z| \\ \langle\downarrow_z| \end{pmatrix} |\ \rangle \qquad (2.8.7)$$

expresses that column as the product of a column of bras with the ket in question. Now, putting things together,

$$|\ \rangle = \left(|\uparrow_z\rangle, |\downarrow_z\rangle\right)\begin{pmatrix} \langle\uparrow_z| \\ \langle\downarrow_z| \end{pmatrix} |\ \rangle \qquad (2.8.8)$$

we learn that

$$\left(|\uparrow_z\rangle, |\downarrow_z\rangle\right)\begin{pmatrix} \langle\uparrow_z| \\ \langle\downarrow_z| \end{pmatrix} = 1 \qquad (2.8.9)$$

because otherwise the statement (2.8.8) cannot be true for all arbitrary kets $|\ \rangle$. The resulting row times column product,

$$|\uparrow_z\rangle\langle\uparrow_z| + |\downarrow_z\rangle\langle\downarrow_z| = 1\,, \qquad (2.8.10)$$

contains "ket times bra" structures that are the abstract generalization of "column times row" products that yield matrices. And just like a matrix acting on a column produces a new column, a "ket-bra" applied to a ket, gives a new ket,

$$\underbrace{\left(|1\rangle\langle 2|\right)}_{\text{ket-bra}} \underbrace{|3\rangle}_{\text{ket}} = \underbrace{|1\rangle}_{\text{ket}} \underbrace{\langle 2|3\rangle}_{\text{bracket}} \qquad (2.8.11)$$

and applied to a bra gives another bra

$$\underbrace{\langle 1|}_{\text{bra}} \underbrace{\left(|2\rangle\langle 3|\right)}_{\text{ket-bra}} = \underbrace{\langle 1|2\rangle}_{\text{bracket}} \underbrace{\langle 3|}_{\text{bra}}\,. \qquad (2.8.12)$$

Splendid, except for the terminology: Rather than of ket-bras, one commonly speaks of *operators*. And then, the unit symbol in (2.8.9) and (2.8.10) is the *identity operator*, that leaves kets and bras unchanged,

$$1| \; \rangle = | \; \rangle, \qquad \langle \; |1 = \langle \; |. \tag{2.8.13}$$

Consistent with the rules about adjoints of kets and bras, such as (2.7.8), we have

$$\left(|1\rangle\langle 2|\right)^\dagger = |2\rangle\langle 1| \tag{2.8.14}$$

as the basic rule for adjoints of ket-bras or operators. This is a particular case of the general statement that

the adjoint of a product is the product
of the adjoints in reverse order.

Here,

$$\left(|1\rangle\langle 2|\right)^\dagger = \langle 2|^\dagger \, |1\rangle^\dagger. \tag{2.8.15}$$

2-11 Show that

$$\langle 1|2\rangle = \langle 2|1\rangle^*$$

must hold for any bra $\langle 1|$ and ket $|2\rangle$ for consistency with (2.8.11) and (2.8.12).

What is true for z kets and bras in (2.8.10) must also hold for x kets and bras,

$$|\!\uparrow_x\rangle\langle\uparrow_x| + |\!\downarrow_x\rangle\langle\downarrow_x| = 1, \tag{2.8.16}$$

which we can verify by translating it into a numerical statement with the aid of the standard rows and columns:

$$\frac{1}{\sqrt 2}\binom{1}{1}\frac{1}{\sqrt 2}(1,1) + \frac{1}{\sqrt 2}\binom{1}{-1}\frac{1}{\sqrt 2}(1,-1)$$
$$= \frac{1}{2}\begin{pmatrix}1&1\\1&1\end{pmatrix} + \frac{1}{2}\begin{pmatrix}1&-1\\-1&1\end{pmatrix} = \begin{pmatrix}1&0\\0&1\end{pmatrix}, \tag{2.8.17}$$

indeed.

2-12 Repeat this for the y kets and bras.

These statements, the one for x above, that for z in (2.8.10) and the one for y of the exercise are examples of *completeness relations* (or *closure relations*), which express that we are not missing something. All possibilities are accounted for: "$+$ in x" *and* "$-$ in x" (or y, or z).

2-13 Consider n pairs of kets, the kth pair denoted by $|a_k\rangle$ and $|b_k\rangle$, that are jointly defined by

$$|a_k\rangle = |\uparrow_z\rangle u_k^* + |\downarrow_z\rangle v_k, \qquad |b_k\rangle = |\uparrow_z\rangle v_k^* - |\downarrow_z\rangle u_k,$$

for $k = 1, 2, \ldots, n$, where the amplitudes u_k and v_k are arbitrary complex numbers. How large are the probabilities $|\langle a_k|b_k\rangle|^2$? How are the probability amplitudes $\langle a_j|a_k\rangle$ and $\langle b_j|b_k\rangle$ related to each other?

2.9 Pauli operators, Pauli matrices

What about other products of kets and bras such as $|\uparrow_z\rangle\langle\downarrow_z|$? This operator represents an apparatus that rejects $\uparrow_z$, but accepts $\downarrow_z$ and turns it into $\uparrow_z$:

$$\left(|\uparrow_z\rangle\langle\downarrow_z|\right)|\uparrow_z\rangle = |\uparrow_z\rangle\underbrace{\langle\downarrow_z|\uparrow_z\rangle}_{=0} = 0,$$

$$\left(|\uparrow_z\rangle\langle\downarrow_z|\right)|\downarrow_z\rangle = |\uparrow_z\rangle\underbrace{\langle\downarrow_z|\downarrow_z\rangle}_{=1} = |\uparrow_z\rangle. \tag{2.9.1}$$

It could be realized by

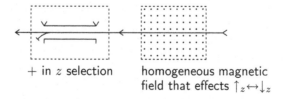

$+$ in z selection $\qquad$ homogeneous magnetic field that effects $\uparrow_z \leftrightarrow \downarrow_z$

for example. And similarly,

$$|\uparrow_z\rangle\langle\downarrow_z| + |\downarrow_z\rangle\langle\uparrow_z| \tag{2.9.2}$$

interchanges $\uparrow_z$ and $\downarrow_z$

$$\Big(|\uparrow_z\rangle\langle\downarrow_z| + |\downarrow_z\rangle\langle\uparrow_z| \Big)|\uparrow_z\rangle = |\downarrow_z\rangle,$$

$$\Big(|\uparrow_z\rangle\langle\downarrow_z| + |\downarrow_z\rangle\langle\uparrow_z| \Big)|\downarrow_z\rangle = |\uparrow_z\rangle, \tag{2.9.3}$$

essentially the action of the homogeneous magnetic field alone. Let us see what it does to x states:

$$\Big(|\uparrow_z\rangle\langle\downarrow_z| + |\downarrow_z\rangle\langle\uparrow_z| \Big)|\uparrow_x\rangle = |\downarrow_z\rangle\frac{1}{\sqrt{2}} + |\uparrow_z\rangle\frac{1}{\sqrt{2}} = |\uparrow_x\rangle, \tag{2.9.4}$$

$$= |\uparrow_z\rangle\tfrac{1}{\sqrt{2}} + |\downarrow_z\rangle\tfrac{1}{\sqrt{2}}$$

that is: $|\uparrow_x\rangle$ is left unaltered;

$$\Big(|\uparrow_z\rangle\langle\downarrow_z| + |\downarrow_z\rangle\langle\uparrow_z| \Big)|\downarrow_x\rangle = |\downarrow_z\rangle\frac{1}{\sqrt{2}} - |\uparrow_z\rangle\frac{1}{\sqrt{2}} = -|\downarrow_x\rangle, \tag{2.9.5}$$

$$= |\uparrow_z\rangle\tfrac{1}{\sqrt{2}} - |\downarrow_z\rangle\tfrac{1}{\sqrt{2}}$$

that is: $|\downarrow_x\rangle$ gets multiplied by -1. Accordingly, it must be true that this operator is also given by the sum $|\uparrow_x\rangle\langle\uparrow_x| - |\downarrow_x\rangle\langle\downarrow_x|$. Indeed, the operator σ_x defined by either one of the expressions

$$\sigma_x = |\uparrow_z\rangle\langle\downarrow_z| + |\downarrow_z\rangle\langle\uparrow_z| = |\uparrow_x\rangle\langle\uparrow_x| - |\downarrow_x\rangle\langle\downarrow_x| \tag{2.9.6}$$

is uniquely defined irrespective of which expression we regard as the definition. We can verify that in a variety of manners, simplest perhaps by involving once more the standard numerical columns and rows:

$$\begin{pmatrix} 0 & 1 \\ 1 & 0 \end{pmatrix} = \underbrace{\begin{pmatrix} 1 \\ 0 \end{pmatrix}(0,1)}_{=\begin{pmatrix} 0 & 1 \\ 0 & 0 \end{pmatrix}} + \underbrace{\begin{pmatrix} 0 \\ 1 \end{pmatrix}(1,0)}_{=\begin{pmatrix} 0 & 0 \\ 1 & 0 \end{pmatrix}}$$

$$= \underbrace{\frac{1}{\sqrt{2}}\begin{pmatrix} 1 \\ 1 \end{pmatrix}\frac{1}{\sqrt{2}}(1,1)}_{=\frac{1}{2}\begin{pmatrix} 1 & 1 \\ 1 & 1 \end{pmatrix}} - \underbrace{\frac{1}{\sqrt{2}}\begin{pmatrix} 1 \\ -1 \end{pmatrix}\frac{1}{\sqrt{2}}(1,-1)}_{=\frac{1}{2}\begin{pmatrix} 1 & -1 \\ -1 & 1 \end{pmatrix}}. \tag{2.9.7}$$

The corresponding operators for y and z are

$$\sigma_y = |\!\uparrow_y\rangle\langle\uparrow_y\!| - |\!\downarrow_y\rangle\langle\downarrow_y\!| \cong \begin{pmatrix} 0 & -i \\ i & 0 \end{pmatrix},$$

$$\sigma_z = |\!\uparrow_z\rangle\langle\uparrow_z\!| - |\!\downarrow_z\rangle\langle\downarrow_z\!| \cong \begin{pmatrix} 1 & 0 \\ 0 & -1 \end{pmatrix}. \tag{2.9.8}$$

In honor of Wolfgang Pauli, the three operators $\sigma_x, \sigma_y, \sigma_z$ are the called *Pauli operators* and their standard matrix representation

$$\sigma_x \cong \begin{pmatrix} 0 & 1 \\ 1 & 0 \end{pmatrix}, \quad \sigma_y \cong \begin{pmatrix} 0 & -i \\ i & 0 \end{pmatrix}, \quad \sigma_z \cong \begin{pmatrix} 1 & 0 \\ 0 & -1 \end{pmatrix} \tag{2.9.9}$$

are known as *Pauli matrices*. The Pauli operators are associated with the three coordinate axes, but since they themselves are arbitrarily oriented, $\sigma_x, \sigma_y, \sigma_z$ must be the cartesian components of a *Pauli vector operator* $\vec{\sigma}$,

$$\vec{\sigma} = \sigma_x \vec{e}_1 + \sigma_y \vec{e}_2 + \sigma_z \vec{e}_3, \tag{2.9.10}$$

a three-dimensional vector with all its typical properties, whose components are operators, however.

By making repeated use of the numerical representation by the Pauli matrices, one easily verifies the fundamental algebraic properties,

unit square: $\quad \sigma_x^2 = 1, \qquad \sigma_y^2 = 1, \qquad \sigma_z^2 = 1;$

cyclic: $\quad \sigma_x\sigma_y = i\sigma_z, \quad \sigma_y\sigma_z = i\sigma_x, \quad \sigma_z\sigma_x = i\sigma_y;$

anticyclic: $\quad \sigma_y\sigma_x = -i\sigma_z, \; \sigma_z\sigma_y = -i\sigma_x, \; \sigma_x\sigma_z = -i\sigma_y. \tag{2.9.11}$

A compact statement that comprises all of these as special cases is

$$\vec{a}\cdot\vec{\sigma}\;\vec{b}\cdot\vec{\sigma} = \vec{a}\cdot\vec{b} + i(\vec{a}\times\vec{b})\cdot\vec{\sigma} \tag{2.9.12}$$

where $\vec{a}$, $\vec{b}$ are arbitrary numerical three-dimensional vectors. Letting $\vec{a}$ and $\vec{b}$ be any of the three unit vectors for the x, y, and z directions, the nine statements of (2.9.11) are recovered. But now we have a formulation of the algebraic properties that makes no reference to a particular choice of cartesian axes. Some immediate consequences are the subject matter of the following exercises.

2-14 Show that $(\vec{a}\cdot\vec{\sigma})^2 = \vec{a}^2$ for any 3-vector $\vec{a}$. What do you get for $\vec{a} = \vec{e}_1 + i\vec{e}_2$?

2-15 Show that $\vec{a}\cdot\vec{\sigma}\;\vec{b}\cdot\vec{\sigma} = -\vec{b}\cdot\vec{\sigma}\;\vec{a}\cdot\vec{\sigma}$ if $\vec{a}$ and $\vec{b}$ are perpendicular to each other.

2-16   Many nonstandard matrix representations supplement the standard representation of (2.9.9), among them

$$\sigma_a \cong \begin{pmatrix} \cos\vartheta & e^{i\varphi_a}\sin\vartheta \\ e^{-i\varphi_a}\sin\vartheta & \cos\vartheta \end{pmatrix} \quad \text{for} \quad a = x, y, z,$$

which is more symmetrical than the standard representation inasmuch as the three matrices for σ_x, σ_y, and σ_z have the same structure and differ only by the respective phase factors $e^{i\varphi_x}$, $e^{i\varphi_y}$, and $e^{i\varphi_z}$. Find the common value of ϑ (with $0 < 2\vartheta < \pi$) and determine a consistent choice for φ_x, φ_y, and φ_z.

2.10   Functions of Pauli operators

Another important consequence is that any function of $\vec{\sigma}$, however complicated, can be regarded as a linear function of $\vec{\sigma}$. That is: if $f(\vec{\sigma})$ is some (operator-valued) function of $\vec{\sigma}$, then it is always possible to find a number a_0 and a 3-vector $\vec{a}$ such that

$$f(\vec{\sigma}) = a_0 + \vec{a} \cdot \vec{\sigma}. \tag{2.10.1}$$

Simple examples are $\sigma_x\sigma_y = i\sigma_z$ and the like, and

$$\underbrace{\sigma_x\sigma_y}_{=-\sigma_y\sigma_x}\sigma_x = -\sigma_y\underbrace{\sigma_x\sigma_x}_{=1} = -\sigma_y \tag{2.10.2}$$

is another. Here we can also argue as indicated by

$$\underbrace{\sigma_x\sigma_y}_{=i\sigma_z}\sigma_x = i\underbrace{\sigma_z\sigma_x}_{=i\sigma_y} = -\sigma_y \tag{2.10.3}$$

with the same outcome, of course. Other examples are found in the following exercises.

2-17   Express $(1+i\sigma_x)\sigma_y(1-i\sigma_x)$ as a linear function of $\vec{\sigma}$.

2-18   Same for $\frac{1}{4}(1+\sigma_x)(1+\sigma_y)(1+\sigma_x)$. Square the outcome.

2-19   If $\vec{n}$ is a unit 3-vector, what is $\left[\frac{1}{2}(1+\vec{n}\cdot\vec{\sigma})\right]^2$?

2-20   What is $\sigma_x\sigma_y\sigma_z$?

2-21 Show that

$$f(\sigma_x)\sigma_z = \sigma_z f(-\sigma_x)$$

holds for the products of Pauli operator σ_z with *any* function of Pauli operator σ_x.

Upon combining, for x states say, the completeness relation

$$\left|\uparrow_x\right\rangle\left\langle\uparrow_x\right| + \left|\downarrow_x\right\rangle\left\langle\downarrow_x\right| = 1 \qquad (2.10.4)$$

with

$$\sigma_x = \left|\uparrow_x\right\rangle\left\langle\uparrow_x\right| - \left|\downarrow_x\right\rangle\left\langle\downarrow_x\right|, \qquad (2.10.5)$$

we find the linear functions of σ_x for "+ in x" selection and "− in x" selection,

$$\frac{1+\sigma_x}{2} = \left|\uparrow_x\right\rangle\left\langle\uparrow_x\right|, \qquad \frac{1-\sigma_x}{2} = \left|\downarrow_x\right\rangle\left\langle\downarrow_x\right|. \qquad (2.10.6)$$

Indeed, they act correctly as projection operators, or projectors, that pick out the respective components when acting on an arbitrary ket:

$$\left|\ \right\rangle = \left|\uparrow_x\right\rangle\alpha + \left|\downarrow_x\right\rangle\beta \quad \text{with} \quad \alpha = \left\langle\uparrow_x\right|\ \rangle, \quad \beta = \left\langle\downarrow_x\right|\ \rangle. \qquad (2.10.7)$$

See

$$\frac{1\pm\sigma_x}{2}\left|\ \right\rangle = \frac{1\pm\sigma_x}{2}\left|\uparrow_x\right\rangle\alpha + \frac{1\pm\sigma_x}{2}\left|\downarrow_x\right\rangle\beta$$

$$= \left|\uparrow_x\right\rangle\frac{1\pm1}{2}\alpha + \left|\downarrow_x\right\rangle\frac{1\mp1}{2}\beta = \begin{cases} \left|\uparrow_x\right\rangle\alpha \\ \left|\downarrow_x\right\rangle\beta \end{cases} \qquad (2.10.8)$$

where we use

$$\sigma_x\left|\uparrow_x\right\rangle = \left|\uparrow_x\right\rangle, \qquad \sigma_x\left|\downarrow_x\right\rangle = -\left|\downarrow_x\right\rangle. \qquad (2.10.9)$$

2-22 Express the operator

$$A = \left|\uparrow_x\right\rangle\left\langle\uparrow_z\right| + \left|\downarrow_x\right\rangle\left\langle\downarrow_z\right|$$

as a linear function of $\vec{\sigma}$. What is A^2?

2.11   Eigenvalues, eigenkets, eigenbras

This particular link between σ_x and the x kets $|{\uparrow_x}\rangle$ and $|{\downarrow_x}\rangle$, namely that the application of σ_x just multiplies by $+1$ or -1, respectively, has the mathematical structure of an *eigenvector equation*. More generally, we have an eigenket equation of the form

$$\underset{\substack{\diagup \quad \diagdown \ \diagup \quad \diagdown \\ \text{operator} \quad \diagdown \ \diagup \quad \text{eigenvalue} \\ \text{eigenket} \\ \text{of } A \text{ to eigenvalue } a}}{A|a\rangle = |a\rangle a} \tag{2.11.1}$$

and

$$\underset{\substack{\diagup \quad \diagdown \ \diagup \quad \diagdown \\ \text{eigenvalue} \quad \diagdown \ \diagup \quad \text{operator} \\ \text{eigenbra} \\ \text{of } A \text{ to eigenvalue } a}}{a\langle a| = \langle a|A} \tag{2.11.2}$$

is the analogous statement about eigenbras, the eigenbra equation.

2-23   Operator A has eigenvalues a_1 and a_2 with eigenbra $\langle a_1|$ and eigenket $|a_2\rangle$. Show that $\langle a_1|a_2\rangle = 0$ if $a_1 \neq a_2$.

2-24   If operator A has an eigenket $|a\rangle$ to eigenvalue a, then $A^\dagger$ has an eigenbra $\langle a^*|$ to eigenvalue a^*. Why? How is $\langle a^*|$ related to $|a\rangle$?

We convince ourselves that there are no other eigenvalues of σ_x than $+1$ and -1. Because if there were another one, denoted by λ, we would need to find amplitudes α and β such that

$$\sigma_x\Big(|{\uparrow_x}\rangle\alpha + |{\downarrow_x}\rangle\beta\Big) = \Big(|{\uparrow_x}\rangle\alpha + |{\downarrow_x}\rangle\beta\Big)\lambda\,, \tag{2.11.3}$$

but the left-hand side is equal to $|{\uparrow_x}\rangle\alpha - |{\downarrow_x}\rangle\beta$, so that

$$\alpha = \lambda\alpha\,, \qquad \beta = -\lambda\beta \tag{2.11.4}$$

have to hold simultaneously. It follows that either $\lambda = 1$, α arbitrary, $\beta = 0$ or $\lambda = -1$, $\alpha = 0$, β arbitrary, which are the cases we know already. The apparent third possibility λ arbitrary, $\alpha = 0$, $\beta = 0$ does not count because then $|\ \rangle = 0$, which is to say that we do not have an eigenket at all.

How does this argument look if we begin with

$$| \rangle = |\uparrow_z\rangle\alpha + |\downarrow_z\rangle\beta \ ? \tag{2.11.5}$$

Then $\sigma_x|\uparrow_z\rangle = |\downarrow_z\rangle$ and $\sigma_x|\downarrow_z\rangle = |\uparrow_z\rangle$ imply

$$\Big(|\uparrow_z\rangle\alpha + |\downarrow_z\rangle\beta\Big)\lambda \underset{\substack{\uparrow \\ \text{(want)}}}{=} \sigma_x\Big(|\uparrow_z\rangle\alpha + |\downarrow_z\rangle\beta\Big) = |\downarrow_z\rangle\alpha + |\uparrow_z\rangle\beta\,, \tag{2.11.6}$$

that is

$$\alpha = \lambda\beta\,, \qquad \beta = \lambda\alpha \tag{2.11.7}$$

or

$$\begin{pmatrix} 0 & 1 \\ 1 & 0 \end{pmatrix}\begin{pmatrix} \alpha \\ \beta \end{pmatrix} = \lambda\begin{pmatrix} \alpha \\ \beta \end{pmatrix} \tag{2.11.8}$$

or

$$\begin{pmatrix} -\lambda & 1 \\ 1 & -\lambda \end{pmatrix}\begin{pmatrix} \alpha \\ \beta \end{pmatrix} = 0\,. \tag{2.11.9}$$

The latter is, of course, just the standard numerical version of

$$(\sigma_x - \lambda)| \rangle = 0\,. \tag{2.11.10}$$

Now, since $| \rangle = 0$, that is: $\alpha = \beta = 0$, is not an option, such a set of coupled linear equations has a solution only if the determinant of the matrix vanishes,

$$0 \underset{\substack{\uparrow \\ \text{(need)}}}{=} \det\left\{\begin{pmatrix} -\lambda & 1 \\ 1 & -\lambda \end{pmatrix}\right\} = \lambda^2 - 1\,. \tag{2.11.11}$$

This implies first that $\lambda = 1$ or $\lambda = -1$ and then $\alpha = \beta$ or $\alpha = -\beta$, respectively, so that

$$\begin{aligned} |\lambda = 1\rangle &\propto |\uparrow_z\rangle + |\downarrow_z\rangle\,, \\ |\lambda = -1\rangle &\propto |\uparrow_z\rangle - |\downarrow_z\rangle\,, \end{aligned} \tag{2.11.12}$$

with the proportionality factors determined by some convention. We usually insist on kets of unit length, so that

$$|\lambda = \pm 1\rangle = \frac{1}{\sqrt{2}}\left(|\uparrow_z\rangle \pm |\downarrow_z\rangle\right) = \begin{cases} |\uparrow_x\rangle, \\ |\downarrow_x\rangle, \end{cases} \qquad (2.11.13)$$

hardly a surprising result. By the same token, we have

$$\sigma_z|\uparrow_z\rangle = |\uparrow_z\rangle, \qquad \sigma_z|\downarrow_z\rangle = -|\downarrow_z\rangle,$$
$$\sigma_y|\uparrow_y\rangle = |\uparrow_y\rangle, \qquad \sigma_y|\downarrow_y\rangle = -|\downarrow_y\rangle, \qquad (2.11.14)$$

and we expect that for *any* component $\vec{e} \cdot \vec{\sigma}$ of Pauli's vector operator $\vec{\sigma}$ there are analogous eigenkets to the same eigenvalues $+1$ and -1.

It is worth checking this in detail, for which purpose we write the *unit vector $\vec{e}$* in its spherical-coordinates parameterization,

$$\vec{e} \cong \begin{pmatrix} \sin\vartheta \cos\varphi \\ \sin\vartheta \sin\varphi \\ \cos\vartheta \end{pmatrix}. \qquad (2.11.15)$$

Then

$$\vec{e} \cdot \vec{\sigma} = \sigma_x \sin\vartheta \cos\varphi + \sigma_y \sin\vartheta \sin\varphi + \sigma_z \cos\vartheta$$
$$\cong \begin{pmatrix} 0 & 1 \\ 1 & 0 \end{pmatrix} \sin\vartheta \cos\varphi + \begin{pmatrix} 0 & -i \\ i & 0 \end{pmatrix} \sin\vartheta \sin\varphi + \begin{pmatrix} 1 & 0 \\ 0 & -1 \end{pmatrix} \cos\vartheta$$
$$= \begin{pmatrix} \cos\vartheta & e^{-i\varphi}\sin\vartheta \\ e^{i\varphi}\sin\vartheta & \cos\vartheta \end{pmatrix} \qquad (2.11.16)$$

once more utilizing the standard numerical representations of σ_x, σ_y, and σ_z. The eigenvector equation

$$(\vec{e} \cdot \vec{\sigma} - \lambda)|\ \rangle = 0 \qquad (2.11.17)$$

then appears as

$$\begin{pmatrix} \cos\vartheta - \lambda & e^{-i\varphi}\sin\vartheta \\ e^{i\varphi}\sin\vartheta & -\cos\vartheta - \lambda \end{pmatrix} \begin{pmatrix} \alpha \\ \beta \end{pmatrix} = 0. \qquad (2.11.18)$$

We evaluate the determinant of this 2×2 matrix,

$$0 = (\cos\vartheta - \lambda)(-\cos\vartheta - \lambda) - (e^{-i\varphi}\sin\vartheta)(e^{i\varphi}\sin\vartheta) = \lambda^2 - 1, \quad (2.11.19)$$
↑
(need)

so that $\lambda = +1$ and $\lambda = -1$ are the eigenvalues, indeed. To find the respective eigenkets, we solve for the amplitudes α, β after choosing one of the eigenvalues. Let us do it for $\lambda = 1$:

$$\begin{pmatrix} \cos\vartheta - 1 & e^{-i\varphi}\sin\vartheta \\ e^{i\varphi}\sin\vartheta & -\cos\vartheta - 1 \end{pmatrix} \begin{pmatrix} \alpha \\ \beta \end{pmatrix} = 0, \quad (2.11.20)$$

or, after introducing $\frac{\vartheta}{2}$,

$$-2 \begin{pmatrix} (\sin\frac{\vartheta}{2})^2 & -e^{-i\varphi}\sin\frac{\vartheta}{2}\cos\frac{\vartheta}{2} \\ -e^{i\varphi}\sin\frac{\vartheta}{2}\cos\frac{\vartheta}{2} & (\cos\frac{\vartheta}{2})^2 \end{pmatrix} \begin{pmatrix} \alpha \\ \beta \end{pmatrix} = 0, \quad (2.11.21)$$

which can also be written as

$$-2 \begin{pmatrix} -e^{-i\varphi}\sin\frac{\vartheta}{2} \\ \cos\frac{\vartheta}{2} \end{pmatrix} \left(-e^{i\varphi}\sin\frac{\vartheta}{2}, \cos\frac{\vartheta}{2} \right) \begin{pmatrix} \alpha \\ \beta \end{pmatrix} = 0. \quad (2.11.22)$$

Now, the column on the left is never $\begin{pmatrix} 0 \\ 0 \end{pmatrix}$, and therefore the product of the central row with the column on the right must vanish, which requires

$$\alpha\, e^{i\varphi}\sin\frac{\vartheta}{2} = \beta\cos\frac{\vartheta}{2} \quad (2.11.23)$$

and implies

$$\begin{pmatrix} \alpha \\ \beta \end{pmatrix} = \begin{pmatrix} \cos\frac{\vartheta}{2} \\ e^{i\varphi}\sin\frac{\vartheta}{2} \end{pmatrix}, \quad (2.11.24)$$

up to an over-all phase factor, which we choose to equal 1. The eigenket to $\lambda = +1$ is thus

$$\left| \uparrow_{\vec{e}} \right\rangle = \left| \uparrow_z \right\rangle \cos\frac{\vartheta}{2} + \left| \downarrow_z \right\rangle e^{i\varphi}\sin\frac{\vartheta}{2}. \quad (2.11.25)$$

Rather than repeating the argument for $\lambda = -1$ we note that we have just looked at

$$(\vec{e} \cdot \vec{\sigma} - 1)\left| \uparrow_{\vec{e}} \right\rangle = 0 \quad (2.11.26)$$

where

$$\vec{e} \cdot \vec{\sigma} - 1 = -2\frac{1 - \vec{e} \cdot \vec{\sigma}}{2} = -2|\downarrow_{\vec{e}}\rangle\langle\downarrow_{\vec{e}}| \qquad (2.11.27)$$

in analogy with $|\downarrow_x\rangle\langle\downarrow_x| = \frac{1}{2}(1 - \sigma_x)$, for example.

Accordingly, we have

$$-2|\downarrow_{\vec{e}}\rangle\langle\downarrow_{\vec{e}}| \,\hat{=}\, -2\begin{pmatrix} -\,e^{-i\varphi}\sin\frac{\vartheta}{2} \\ \cos\frac{\vartheta}{2} \end{pmatrix}\left(-\,e^{i\varphi}\sin\frac{\vartheta}{2},\ \cos\frac{\vartheta}{2}\right) \qquad (2.11.28)$$

and read off that

$$|\downarrow_{\vec{e}}\rangle \,\hat{=}\, \begin{pmatrix} -\,e^{-i\varphi}\sin\frac{\vartheta}{2} \\ \cos\frac{\vartheta}{2} \end{pmatrix} \qquad (2.11.29)$$

or

$$|\downarrow_{\vec{e}}\rangle = |\uparrow_z\rangle\left(-\,e^{-i\varphi}\sin\frac{\vartheta}{2}\right) + |\downarrow_z\rangle\cos\frac{\vartheta}{2}\,. \qquad (2.11.30)$$

It is a matter of inspection to verify that

$$\vec{e} \cdot \vec{\sigma}|\uparrow_{\vec{e}}\rangle = |\uparrow_{\vec{e}}\rangle \quad\text{and}\quad \vec{e} \cdot \vec{\sigma}|\downarrow_{\vec{e}}\rangle = -|\downarrow_{\vec{e}}\rangle \qquad (2.11.31)$$

and that the orthogonality relation

$$\langle\uparrow_{\vec{e}}|\downarrow_{\vec{e}}\rangle = 0 \qquad (2.11.32)$$

holds. Also, we have built in the normalization,

$$\langle\uparrow_{\vec{e}}|\uparrow_{\vec{e}}\rangle = 1\,, \qquad \langle\downarrow_{\vec{e}}|\downarrow_{\vec{e}}\rangle = 1\,, \qquad (2.11.33)$$

and the generalization of the completeness relations (2.8.10) and (2.8.16) is

$$|\uparrow_{\vec{e}}\rangle\langle\uparrow_{\vec{e}}| + |\downarrow_{\vec{e}}\rangle\langle\downarrow_{\vec{e}}| = 1\,. \qquad (2.11.34)$$

2-25 Verify (2.11.32), (2.11.33), and (2.11.34).

2-26 For $\vartheta = \frac{\pi}{2}, \varphi = 0$ you have $\vec{e} \cdot \vec{\sigma} = \sigma_x$. Compare the kets $|\uparrow_{\vec{e}}\rangle$, $|\downarrow_{\vec{e}}\rangle$ with the kets $|\uparrow_x\rangle$, $|\downarrow_x\rangle$ found previously. Repeat for $\vartheta = \frac{\pi}{2}, \varphi = \frac{\pi}{2}$ when $\vec{e} \cdot \vec{\sigma} = \sigma_y$.

2.12 Wave-particle duality

All this rather simple mathematics may not be so impressive, but there really is a fundamental physical insight in it. There is a *continuum* of $\vec{e} \cdot \vec{\sigma}$ operators, each referring to magnetic properties of the silver atoms associated with the specified direction $\vec{e}$ in space, but for each direction the values are *discrete*, invariably $+1$ and -1.

This union of the discrete and the continuous is at the heart of quantum mechanics. Inasmuch as discrete features are associated with particle-like objects in classical physics (planets in orbit, falling stones, ...) whereas continuous features are associated with wave phenomena (sound waves, radio waves, ...), the manifestation of both in the same class of phenomena triggered the historical debate about *wave-particle duality*, commencing with Albert Einstein's paper of 1905 on the photoelectric effect, but still continuing today.

The present example of continuous directional options with discrete alternatives for each of them is the paradigmatic illustration of "waves and particles in peaceful coexistence" or the discrete-and-continuous. And please do not fail to note how this came about as a natural, unavoidable consequence of the experimental fact that atoms are deflected up *or* down in a Stern–Gerlach experiment, with no other option available to them.

2.13 Expectation value

We observed above that the Pauli operators are of the form

$$
\left. \begin{array}{c} \sigma_x \\ \sigma_y \\ \sigma_z \end{array} \right\} = \left\{ \begin{array}{c} |{\uparrow}_x\rangle\langle{\uparrow}_x| - |{\downarrow}_x\rangle\langle{\downarrow}_x| \\ |{\uparrow}_y\rangle\langle{\uparrow}_y| - |{\downarrow}_y\rangle\langle{\downarrow}_y| \\ |{\uparrow}_z\rangle\langle{\uparrow}_z| - |{\downarrow}_z\rangle\langle{\downarrow}_z| \end{array} \right\}
$$

$$
= (+1)(+\text{selector}) + (-1)(-\text{selector}) \qquad (2.13.1)
$$

and noted that the numbers $+1$ and -1 appearing here are the eigenvalues of these operators. We have, in effect, introduced a mathematical symbol that represents the physical property measured by the respective Stern–Gerlach apparatus, with the understanding that we assign $+1$ to "deflected

up" and −1 to "deflected down",

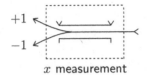

x measurement

While this is the conventional, and perhaps most natural, assignment of numbers that characterize the measurement result, it is, of course by no means unique. Just as well we could have chosen any other two numbers, real or complex. If we denote them by λ_+ and λ_-, we have this situation:

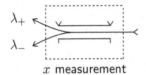

x measurement

and the corresponding symbol is

$$\Lambda \equiv |{\uparrow_x}\rangle\lambda_+\langle{\uparrow_x}| + |{\downarrow_x}\rangle\lambda_-\langle{\downarrow_x}| \tag{2.13.2}$$

or, if we recall that $\frac{1}{2}(1 \pm \sigma_x)$ are the projectors in question,

$$\Lambda = \lambda_+\frac{1+\sigma_x}{2} + \lambda_-\frac{1-\sigma_x}{2} = \frac{1}{2}(\lambda_+ + \lambda_-) + \frac{1}{2}(\lambda_+ - \lambda_-)\sigma_x. \tag{2.13.3}$$

The kets $|{\uparrow_x}\rangle$ and $|{\downarrow_x}\rangle$ remain eigenkets, but the eigenvalues are now λ_+ and λ_-,

$$\Lambda|{\uparrow_x}\rangle = |{\uparrow_x}\rangle\lambda_+\,, \qquad \Lambda|{\downarrow_x}\rangle = |{\downarrow_x}\rangle\lambda_-\,. \tag{2.13.4}$$

There is nothing in the physics of the situation that would dictate one particular choice for λ_+ and λ_-, and so all Λ operators constructed in this way refer to the magnetic property of the silver atoms measured by a Stern–Gerlach apparatus for x direction. Put differently, an apparatus that measures σ_x measures at the same time all functions of σ_x — recall here that the most general function is a constant added to a multiple of σ_x, a linear function of σ_x.

Now, having opted for a particular λ_+ and λ_- pair, the probability of getting outcome λ_+ is just the probability for up-deflection,

$$\mathrm{prob}(\lambda_+) = \left|\langle{\uparrow_x}|\ \rangle\right|^2, \tag{2.13.5}$$

and likewise

$$\text{prob}(\lambda_-) = \left|\langle\downarrow_x|\ \rangle\right|^2 \tag{2.13.6}$$

for λ_- and down-deflection.

An individual atom will either be recorded with value λ_+ or value λ_-, and these probabilities state the relative frequencies with which the values occur if very many atoms are measured. The average value,

$$\lambda_+\text{prob}(\lambda_+) + \lambda_-\text{prob}(\lambda_-), \tag{2.13.7}$$

is the weighted sum of the two measurement results. Introducing the probabilities of (2.13.5) and (2.13.6), we have

$$\begin{aligned}
\lambda_+\left|\langle\uparrow_x|\ \rangle\right|^2 + \lambda_-\left|\langle\downarrow_x|\ \rangle\right|^2 &= \langle\ |\uparrow_x\rangle\lambda_+\langle\uparrow_x|\ \rangle + \langle\ |\downarrow_x\rangle\lambda_-\langle\downarrow_x|\ \rangle \\
&= \langle\ |\Big(|\uparrow_x\rangle\lambda_+\langle\uparrow_x| + |\downarrow_x\rangle\lambda_-\langle\downarrow_x|\Big)|\ \rangle \\
&= \langle\ |\Lambda|\ \rangle.
\end{aligned} \tag{2.13.8}$$

So, the eigenvalues of Λ are the measurement results, and their statistical mean value is the number that we obtain by sandwiching Λ by the bra $\langle\ |$ and the ket $|\ \rangle$ that describe the atoms that are measured. A common shorthand notation is $\langle\Lambda\rangle$ for such an average,

$$\langle\Lambda\rangle = \lambda_+\text{prob}(\lambda_+) + \lambda_-\text{prob}(\lambda_-) = \langle\ |\Lambda|\ \rangle, \tag{2.13.9}$$

and standard terminology is to call it the "mean value of Λ" or the "*expectation value of Λ*". The latter is very widespread, almost universal terminology. But note that we are not talking about an "expected value". Rather, we expect to get either λ_+ or λ_-, that is: one of the eigenvalues of Λ, in each measurement act and the statistical average of those numbers is the expectation value of Λ.

Probabilities are expectation values as well:

$$\begin{aligned}
\text{prob}(\lambda_+) = \left|\langle\uparrow_x|\ \rangle\right|^2 &= \langle\ |\Big(|\uparrow_x\rangle\langle\uparrow_x|\Big)|\ \rangle \\
&= \langle\ |\frac{1+\sigma_x}{2}|\ \rangle \\
&= \left\langle\frac{1+\sigma_x}{2}\right\rangle,
\end{aligned} \tag{2.13.10}$$

and likewise

$$\text{prob}(\lambda_-) = \left\langle\frac{1-\sigma_x}{2}\right\rangle. \tag{2.13.11}$$

With this in mind, it should be clear that *all* statistical measurement results are expectation values of the appropriate operators. Probabilities, in particular, are the expectation values of the operators that select the outcome, such as $\frac{1}{2}(1 \pm \sigma_x)$ in our example.

2-27 For the situation considered in Section 2.5, where the ket of (2.7.6) applies, show that

$$\langle \sigma_x \rangle = 2 \operatorname{Re}(\alpha^* \beta) \;, \quad \langle \sigma_y \rangle = 2 \operatorname{Im}(\alpha^* \beta) \;, \quad \langle \sigma_z \rangle = |\alpha|^2 - |\beta|^2 \;.$$

What is their significance in the context of the measurement sketched by the figure on page 31?

2.14   Trace

To proceed further, we need to introduce a new mathematical operation, called the *trace*. It assigns a number to each operator, the basic relation being

$$\operatorname{tr}\left\{|1\rangle\langle 2|\right\} = \langle 2|1\rangle \,, \tag{2.14.1}$$

pronounced: the trace of $|1\rangle\langle 2|$ is $\langle 2|1\rangle$. Since it is a simple number, this is not a reversible mathematical operation: Knowing the operator, you can work out the trace, but from the knowledge of the trace alone, you cannot infer the operator uniquely. The trace is a *linear operation*,

$$\operatorname{tr}\{X + Y\} = \operatorname{tr}\{X\} + \operatorname{tr}\{Y\} \;,$$
$$\operatorname{tr}\{\lambda X\} = \lambda \operatorname{tr}\{X\} \;, \tag{2.14.2}$$

for all operators X, Y and all complex numbers λ. This linearity is necessary for consistency with the linear structure of the ket and bra spaces, as illustrated by

$$|1\rangle = |3\rangle\alpha + |4\rangle\beta \tag{2.14.3}$$

and

$$\begin{aligned} \operatorname{tr}\left\{|1\rangle\langle 2|\right\} = \langle 2|1\rangle &= \langle 2|3\rangle\alpha + \langle 2|4\rangle\beta \\ &= \alpha \operatorname{tr}\left\{|3\rangle\langle 2|\right\} + \beta \operatorname{tr}\left\{|4\rangle\langle 2|\right\} \end{aligned} \tag{2.14.4}$$

and the like.

2-28 Evaluate $\operatorname{tr}\left\{|{\uparrow_x}\rangle\langle{\downarrow_y}|\right\}$ and $\operatorname{tr}\{1\} = \operatorname{tr}\left\{|{\uparrow_x}\rangle\langle{\uparrow_x}| + |{\downarrow_x}\rangle\langle{\downarrow_x}|\right\}$.

For an arbitrary operator X, we can use *some* decomposition of the identity,

$$|\uparrow\rangle\langle\uparrow| + |\downarrow\rangle\langle\downarrow| = 1,\qquad (2.14.5)$$

to establish

$$\begin{aligned}
\mathrm{tr}\{X\} &= \mathrm{tr}\left\{\left(|\uparrow\rangle\langle\uparrow| + |\downarrow\rangle\langle\downarrow|\right)X\right\}\\
&= \mathrm{tr}\left\{|\uparrow\rangle\left(\langle\uparrow|X\right) + |\downarrow\rangle\left(\langle\downarrow|X\right)\right\}\\
&= \langle\uparrow|X|\uparrow\rangle + \langle\downarrow|X|\downarrow\rangle,\qquad (2.14.6)
\end{aligned}$$

which says the following: If we represent X by a 2×2 matrix in accordance with the usual procedure,

$$\begin{aligned}
X = 1X1 &= \left(|\uparrow\rangle\langle\uparrow| + |\downarrow\rangle\langle\downarrow|\right)X\left(|\uparrow\rangle\langle\uparrow| + |\downarrow\rangle\langle\downarrow|\right)\\
&= \underbrace{\left(|\uparrow\rangle, |\downarrow\rangle\right)}_{\text{row of kets}}\underbrace{\begin{pmatrix}\langle\uparrow|X|\uparrow\rangle & \langle\uparrow|X|\downarrow\rangle\\ \langle\downarrow|X|\uparrow\rangle & \langle\downarrow|X|\downarrow\rangle\end{pmatrix}}_{2\times 2\text{ matrix}}\underbrace{\begin{pmatrix}\langle\uparrow|\\ \langle\downarrow|\end{pmatrix}}_{\text{column of bras}}\qquad (2.14.7)
\end{aligned}$$

or

$$X \mathrel{\widehat{=}} \begin{pmatrix}\langle\uparrow|X|\uparrow\rangle & \langle\uparrow|X|\downarrow\rangle\\ \langle\downarrow|X|\uparrow\rangle & \langle\downarrow|X|\downarrow\rangle\end{pmatrix}\qquad (2.14.8)$$

for short, then $\mathrm{tr}\{X\}$ is simply the sum of the diagonal matrix elements. Note that this statement is true irrespective of whether the reference states are up/down in x, or y, or z, or any other direction.

2-29   Evaluate the trace of $3+\sigma_x+\sigma_z$ by using the matrix representations referring to x, y, and z.

In addition to its linearity, the trace operation has another important property, namely that

$$\mathrm{tr}\{XY\} = \mathrm{tr}\{YX\}\qquad (2.14.9)$$

for any two operators X and Y. In view of the linearity, it suffices to consider $X = |1\rangle\langle 2|$ and $Y = |3\rangle\langle 4|$,

$$
\begin{aligned}
\mathrm{tr}\{XY\} = \mathrm{tr}\{|1\rangle\langle 2|3\rangle\langle 4|\} &= \langle 2|3\rangle\mathrm{tr}\{|1\rangle\langle 4|\} \\
&= \langle 2|3\rangle\langle 4|1\rangle \\
&= \langle 4|1\rangle\mathrm{tr}\{|3\rangle\langle 2|\} \\
&= \mathrm{tr}\{|3\rangle\langle 4|1\rangle\langle 2|\} \\
&= \mathrm{tr}\{YX\}, \quad\quad\quad\quad (2.14.10)
\end{aligned}
$$

indeed.

As an immediate consequence, we note the *cyclic property of the trace*,

$$
\mathrm{tr}\{XYZ\} = \mathrm{tr}\{YZX\} = \mathrm{tr}\{ZXY\} \quad\quad\quad\quad (2.14.11)
$$

for any three operators X, Y, and Z, and analogous statements about the cyclic permutation of factors hold for four, five, or more factors.

Now, having added the trace operation to our toolbox, let us reconsider the expectation value of operator Λ,

$$
\langle \Lambda \rangle = \langle \ |\Lambda| \ \rangle = \langle \ |\big(\Lambda| \ \rangle\big) = \mathrm{tr}\Big\{\big(\Lambda| \ \rangle\big)\langle \ |\Big\} \quad\quad\quad\quad (2.14.12)
$$

that is:

$$
\langle \Lambda \rangle = \mathrm{tr}\{\Lambda| \ \rangle\langle \ |\} \quad\quad\quad\quad (2.14.13)
$$

where we succeeded in separating the information about the atoms measured from the information about the measured property.

2-30   Operator Λ has eigenvalues λ_1 and λ_2. Show that their sum is the trace, $\mathrm{tr}\{\Lambda\} = \lambda_1 + \lambda_2$.

2-31   Which matrices represent $\sigma_x + \sigma_y + \sigma_z$ in the standard representation of (2.9.9) and the nonstandard representation of Exercise 2-16 on page 44? Verify that the two matrices have the same trace and the same determinant. Use their values to find the eigenvalues of $\sigma_x + \sigma_y + \sigma_z$.

2.15   Statistical operator

To go beyond mere rewriting of previously obtained expressions, we get to something new by asking the question: How do we describe atoms prepared by a mixing procedure, such as 50% as "+ in x" and 50% as "+ in z"? We answer the question by looking at the resulting expectation value of Λ:

$$\text{atoms ``+ in } x\text{'' : } \langle\Lambda\rangle = \text{tr}\{\Lambda|\uparrow_x\rangle\langle\uparrow_x|\} \ ,$$

$$\text{atoms ``+ in } z\text{'' : } \langle\Lambda\rangle = \text{tr}\{\Lambda|\uparrow_z\rangle\langle\uparrow_z|\} \ ,$$

$$50\% \text{ of each kind : } \langle\Lambda\rangle = \frac{1}{2}\text{tr}\{\Lambda|\uparrow_x\rangle\langle\uparrow_x|\} + \frac{1}{2}\text{tr}\{\Lambda|\uparrow_z\rangle\langle\uparrow_z|\}$$

$$= \text{tr}\left\{\Lambda\left(\frac{1}{2}|\uparrow_x\rangle\langle\uparrow_x| + \frac{1}{2}|\uparrow_z\rangle\langle\uparrow_z|\right)\right\} \quad (2.15.1)$$

or

$$\langle\Lambda\rangle = \text{tr}\{\Lambda\rho\} \quad\quad\quad (2.15.2)$$

with

$$\rho = \frac{1}{2}\left(|\uparrow_x\rangle\langle\uparrow_x| + |\uparrow_z\rangle\langle\uparrow_z|\right). \quad\quad\quad (2.15.3)$$

This new mathematical object ρ, which summarizes all we know about the preparation of the atoms is called *statistical operator* or *state operator* (sometimes "probability operator") and, somewhat misleadingly but rather commonly, *density matrix*. Our preferred terminology is *statistical operator*, or simply *state*.

As all operators, also the statistical operator is a function of Pauli's vector operator $\vec{\sigma}$, a linear function in fact. For the example above, we have

$$\rho = \frac{1}{2}\frac{1+\sigma_x}{2} + \frac{1}{2}\frac{1+\sigma_z}{2} = \frac{1}{2}\left(1 + \frac{1}{2}\sigma_x + \frac{1}{2}\sigma_z\right), \quad\quad (2.15.4)$$

and in general we get

$$\rho = a_0 + \vec{a}\cdot\vec{\sigma} \quad\quad\quad (2.15.5)$$

with a numerical vector $\vec{a}$ and a number a_0 (a multiple of the identity). The probabilities associated with an arbitrary direction are the expectation values of $|\uparrow\rangle\langle\uparrow|$ and $|\downarrow\rangle\langle\downarrow|$,

$$\text{prob(``up'')} = \text{tr}\{|\uparrow\rangle\langle\uparrow|\rho\} = \langle\uparrow|\rho|\uparrow\rangle,$$

$$\text{prob(``down'')} = \text{tr}\{|\downarrow\rangle\langle\downarrow|\rho\} = \langle\downarrow|\rho|\downarrow\rangle, \quad\quad (2.15.6)$$

and their unit sum

$$1 = \text{prob}(\text{``up''}) + \text{prob}(\text{``down''})$$
$$= \text{tr}\Big\{ \underbrace{\big(|{\uparrow}\rangle\langle{\uparrow}| + |{\downarrow}\rangle\langle{\downarrow}|\big)}_{=1} \rho \Big\} = \text{tr}\{\rho\}\,, \qquad (2.15.7)$$

or

$$1 = \langle{\uparrow}|\rho|{\uparrow}\rangle + \langle{\downarrow}|\rho|{\downarrow}\rangle = \text{tr}\{\rho\}\,, \qquad (2.15.8)$$

states

$$\text{tr}\{\rho\} = 1 \qquad (2.15.9)$$

as the condition for unit total probability. Now, since the Pauli operators $\sigma_x, \sigma_y, \sigma_z$ are traceless,

$$\text{tr}\{\sigma_x\} = 0\,, \qquad \text{tr}\{\sigma_y\} = 0\,, \qquad \text{tr}\{\sigma_z\} = 0\,, \qquad (2.15.10)$$

which is easily verified by a look at the diagonal sum of their standard matrix representations in (2.9.9), and the trace of the identity operator is 2,

$$\text{tr}\{1\} = 2\,, \qquad (2.15.11)$$

we obtain

$$\text{tr}\{\rho\} = 2a_0 = 1 \quad \text{or} \quad a_0 = \frac{1}{2}\,. \qquad (2.15.12)$$

Further consider

$$\langle\sigma_x\rangle = \text{tr}\{\sigma_x\rho\} = \text{tr}\{\sigma_x(a_0 + \vec{a}\cdot\vec{\sigma})\}$$
$$= a_0\underbrace{\text{tr}\{\sigma_x\}}_{=0} + \text{tr}\{a_x + \underbrace{a_y\text{i}\sigma_z}_{\to 0} + \underbrace{a_z(-\text{i}\sigma_y)}_{\to 0}\} \qquad (2.15.13)$$

so that $\langle\sigma_x\rangle = 2a_x$ and likewise $\langle\sigma_y\rangle = 2a_y$, $\langle\sigma_z\rangle = 2a_z$, compactly summarized in

$$\langle\vec{\sigma}\rangle = 2\vec{a}\,. \qquad (2.15.14)$$

Thus the general statistical operator is of the form

$$\rho = \frac{1}{2}(1 + \vec{s}\cdot\vec{\sigma}) \quad \text{with} \quad \vec{s} = \langle\vec{\sigma}\rangle = \text{tr}\{\vec{\sigma}\rho\}\,. \qquad (2.15.15)$$

What then are the probabilities for "up" and "down" in an arbitrary direction specified by unit vector $\vec{e}$? Since the respective projectors are

$$|\uparrow\rangle\langle\uparrow| = \frac{1}{2}(1 + \vec{e}\cdot\vec{\sigma}),$$

$$|\downarrow\rangle\langle\downarrow| = \frac{1}{2}(1 - \vec{e}\cdot\vec{\sigma}), \tag{2.15.16}$$

we have

$$\left.\begin{array}{r} \text{prob("up")} \\ \text{prob("down")} \end{array}\right\} = \mathrm{tr}\left\{\frac{1}{2}(1 \pm \vec{e}\cdot\vec{\sigma})\rho\right\}$$

$$= \mathrm{tr}\left\{\frac{1}{2}(1 \pm \vec{e}\cdot\vec{\sigma})\frac{1}{2}(1 + \vec{s}\cdot\vec{\sigma})\right\}$$

$$= \mathrm{tr}\left\{\frac{1}{4} \pm \frac{1}{4}\vec{e}\cdot\vec{\sigma} + \frac{1}{4}\vec{s}\cdot\vec{\sigma} \pm \frac{1}{4}\vec{e}\cdot\vec{\sigma}\,\vec{s}\cdot\vec{\sigma}\right\}, \tag{2.15.17}$$

that is

$$\left.\begin{array}{r} \text{prob("up")} \\ \text{prob("down")} \end{array}\right\} = \frac{1}{2} \pm \frac{1}{4}\mathrm{tr}\{\vec{e}\cdot\vec{\sigma}\,\vec{s}\cdot\vec{\sigma}\}. \tag{2.15.18}$$

We evaluate the remaining trace by first recalling that

$$\vec{e}\cdot\vec{\sigma}\,\vec{s}\cdot\vec{\sigma} = \vec{e}\cdot\vec{s} + \mathrm{i}(\vec{e}\times\vec{s})\cdot\vec{\sigma}, \tag{2.15.19}$$

see (2.9.12), and noting once more that $\mathrm{tr}\{1\} = 2, \mathrm{tr}\{\vec{\sigma}\} = 0$, with the consequence

$$\left.\begin{array}{r} \text{prob("up")} \\ \text{prob("down")} \end{array}\right\} = \frac{1}{2} \pm \frac{1}{2}\vec{e}\cdot\vec{s} = \frac{1}{2}(1 \pm \vec{e}\cdot\vec{s}). \tag{2.15.20}$$

Both probabilities are numbers in the range $0 \cdots 1$ so that

$$-1 \le \vec{e}\cdot\vec{s} \le 1 \tag{2.15.21}$$

for all unit vectors $\vec{e}$ and all permissible vectors $\vec{s}$. It follows that the length of $\vec{s}$ cannot exceed unity,

$$|\vec{s}| = \sqrt{\vec{s}\cdot\vec{s}} \le 1. \tag{2.15.22}$$

2-32 A source emits atoms such that each of them is either "+ in x" or "− in z", choosing randomly between these options, with equal chances for both. What are the probabilities of (i) finding the next atom as "+ in x" or "− in x"; (ii) finding the next atom as "+ in y" or "− in y"; (iii) finding the next atom as "+ in z" or "− in z", when the respective experiments are performed?

2-33 Show that $\mathrm{tr}\{\rho^2\} \leq 1$. When does "=" apply?

2-34 We define four hermitian operators $W_0, \ldots, W_3$ by

$$W_0 = \frac{1}{2}(1 + \sigma_x + \sigma_y + \sigma_z),$$
$$W_1 = \sigma_x W_0 \sigma_x, \quad W_2 = \sigma_y W_0 \sigma_y, \quad W_3 = \sigma_z W_0 \sigma_z.$$

Show that they are normalized to unit trace, $\mathrm{tr}\{W_j\} = 1$ for $j = 0, \ldots, 3$, and mutually orthogonal in the sense of

$$\mathrm{tr}\{W_j W_k\} = 2\delta_{jk} \quad \text{for } j, k = 0, \ldots, 3.$$

2-35 Show that these W_js are also complete, that is: any operator Λ can be written as a weighted sum of the W_js,

$$\Lambda = \frac{1}{2}\sum_{j=0}^{3} \lambda_j W_j \quad \text{with } \lambda_j = \mathrm{tr}\{\Lambda W_j\}.$$

Express $\mathrm{tr}\{\Lambda\}$ and $\mathrm{tr}\{\Lambda^\dagger \Lambda\}$ in terms of the coefficients λ_j.

2-36 If the statistical operator is $\rho = \frac{1}{2}\sum_k r_k W_k$, what is the expectation value $\langle \Lambda \rangle$ in terms of the coefficients λ_j and r_k?

2.16   Mixtures and blends

For the example of (2.15.4), we have $\vec{s} = \frac{1}{2}(\vec{e}_1 + \vec{e}_3) \,\hat{=}\, (\frac{1}{2}, 0, \frac{1}{2})$. We can also write this ρ as

$$\rho = \frac{3 + \sqrt{3}}{6} \frac{1 + \left(\frac{\sqrt{3}}{2}\sigma_x + \frac{1}{2}\sigma_z\right)}{2} + \frac{3 - \sqrt{3}}{6} \frac{1 + \left(-\frac{\sqrt{3}}{2}\sigma_x + \frac{1}{2}\sigma_z\right)}{2}, \quad (2.16.1)$$

which seems to tell us that ρ is obtained by putting together a fraction of $(3 + \sqrt{3})/6 = 78.9\%$ of atoms that are up in the direction of the unit vector with coordinates $\frac{1}{2}(\sqrt{3}, 0, 1)$ and a fraction of $(3 - \sqrt{3})/6 = 21.1\%$

of atoms that are up in the direction of the unit vector with coordinates $\frac{1}{2}(-\sqrt{3}, 0, 1)$. It is the same statistical operator ρ in both cases, although the preparation procedures are quite different. And, clearly, if you have two different ways of blending one and the same mixture, you have many.

The last sentence introduces some terminology. In expressions such as

$$\rho = \frac{1}{2}\left(1 + \frac{1}{2}\sigma_x + \frac{1}{2}\sigma_z\right) = \frac{1}{2}\frac{1 + \sigma_x}{2} + \frac{1}{2}\frac{1 + \sigma_z}{2} \qquad (2.16.2)$$

we call the weighted sum of projectors on the right a *blend* of the *mixture* on the left. As we have seen, there are many ways, as a rule, to blend any given mixture. Can you tell, by looking at the mixture, how it was blended from ingredients? No, you cannot, because all statistical data you can get are expectation values,

$$\langle \Lambda \rangle = \text{tr}\{\Lambda\rho\} , \qquad (2.16.3)$$

and for them only the mixture ρ is relevant, not how it is blended. In the example above, it is all right to say that the mixture is "as if 50% are $\uparrow_x$ and 50% are $\uparrow_z$", but this is only one *as-if reality* of many.

2-37 Atoms are prepared by blending $\frac{1}{2}$ of $\uparrow_x$ with $\frac{1}{3}$ of $\uparrow_y$ and $\frac{1}{6}$ of $\downarrow_z$. What is the resulting statistical operator?

2-38 For this preparation, what is the probability of "up" in direction $\vec{e} = \frac{1}{3}(2\,\vec{e}_1 - \vec{e}_2 + 2\,\vec{e}_3)$?

2.17 Nonselective measurement

Having statistical operators at our disposal for the characterization of approaching atoms, we can finally close a gap, namely how do we describe, in the mathematical formalism, the procedure of measuring σ_z, say, without however selecting a partial beam? The situation is this:

SG apparatus
for the z direction

Clearly, the emerging atoms are here blended from $\uparrow_z$ and $\downarrow_z$, and the weights are the probabilities for an incoming atom to be deflected correspondingly. Accordingly, we have

$$\rho_{\text{out}} = \underbrace{\frac{1}{2}\left(1 + s_z\right)}_{\substack{\text{probability} \\ \text{for } \uparrow_z}} \underbrace{\frac{1}{2}\left(1 + \sigma_z\right)}_{\substack{\text{projector} \\ \text{on } \uparrow_z}} + \underbrace{\frac{1}{2}\left(1 - s_z\right)}_{\substack{\text{probability} \\ \text{for } \downarrow_z}} \underbrace{\frac{1}{2}\left(1 - \sigma_z\right)}_{\substack{\text{projector} \\ \text{on } \downarrow_z}} \qquad (2.17.1)$$

or

$$\rho_{\text{out}} = \frac{1}{2}\left(1 + s_z\sigma_z\right). \qquad (2.17.2)$$

The comparison with

$$\rho_{\text{in}} = \frac{1}{2}\left(1 + s_x\sigma_x + s_y\sigma_y + s_z\sigma_z\right) \qquad (2.17.3)$$

shows that the nonselective measurement effectively wipes out the components referring to the perpendicular directions:

$$\vec{s} \mathrel{\widehat{=}} \underbrace{\left(s_x,\, s_y,\, s_z\right)}_{\text{in}} \longrightarrow \vec{s} \mathrel{\widehat{=}} \underbrace{\left(0,\, 0,\, s_z\right)}_{\text{out}}. \qquad (2.17.4)$$

It does not matter, of course, whether we (or someone else) cares to keep a record which atom was deflected which way, because this information is not taken into account anyway. What enters into the statistical operator is solely the information upon which we condition our predictions. In the above example, this is: Provided that the source emits atoms in state ρ_{in} and that they traverse a z-measuring Stern–Gerlach apparatus without being selected, what is then the expectation value of Λ for the beam thus prepared? Answer:

$$\langle \Lambda \rangle = \text{tr}\{\Lambda \rho_{\text{out}}\} = \frac{1}{2}\text{tr}\{\Lambda\} + \frac{1}{2}s_z\text{tr}\{\Lambda\sigma_z\}. \qquad (2.17.5)$$

2-39 Show that the net effect of the nonselective z measurement can be described by

$$\rho_{\text{in}} \to \rho_{\text{out}} = \frac{1}{2}\left(\rho_{\text{in}} + \sigma_z\rho_{\text{in}}\sigma_z\right).$$

How would you describe the effect of a nonselective x measurement?

2.18   Entangled atom pairs

So far we have been dealing exclusively with the quantum aspects of the magnetic properties of simple silver atoms. Let us now move on and turn to more complicated systems. For a start, we consider pairs of silver atoms in a situation much like that discussed in Section 1.2 for paired photons, that is:

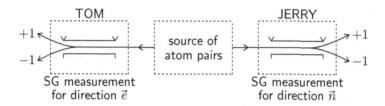

The source emits pairs of atoms, one traveling to the left, the other to the right. On the left, where *Tom* is taking data, his atom passes through a Stern–Gerlach magnet, set for probing direction $\vec{e}$ and number $+1$ or -1 is recorded for each atom. Likewise on the right, where *Jerry* orients his apparatus to probe direction $\vec{n}$ and records $+1$ or -1 for each atom.

For a particular choice of direction, they eventually calculate the Bell correlation $C(\vec{e}, \vec{n})$ of (1.2.1) in accordance with

$$C(\vec{e}, \vec{n}) = \text{(relative frequency of products} = +1)$$
$$- \text{(relative frequency of products} = -1), \quad (2.18.1)$$

or expressed in terms of probabilities for $\uparrow\uparrow$, $\uparrow\downarrow$, $\downarrow\uparrow$, $\downarrow\downarrow$,

$$C(\vec{e}, \vec{n}) = \text{prob}(\uparrow\uparrow \text{ or } \downarrow\downarrow) - \text{prob}(\uparrow\downarrow \text{ or } \downarrow\uparrow). \quad (2.18.2)$$

Clearly, the value of $C(\vec{e}, \vec{n})$ depends on the directions that are probed, and also on the two-atom state emitted by the source.

The choice of directions enters the Bell correlation $C(\vec{e}, \vec{n})$ in the form of projection operators $\frac{1}{2}(1 \pm \vec{e} \cdot \vec{\sigma}^{(1)})$ and $\frac{1}{2}(1 \pm \vec{n} \cdot \vec{\sigma}^{(2)})$ where $\vec{\sigma}^{(1)}$ is the Pauli vector operator for Tom's atom on the left, and $\vec{\sigma}^{(2)}$ is that for Jerry's atom on the right. The respective probabilities are then given by the expectation values of the corresponding products of projection operators,

$$\text{prob}(\uparrow\uparrow) = \left\langle \frac{1 + \vec{e} \cdot \vec{\sigma}^{(1)}}{2} \frac{1 + \vec{n} \cdot \vec{\sigma}^{(2)}}{2} \right\rangle ,$$

$$\text{prob}(\downarrow\downarrow) = \left\langle \frac{1 - \vec{e} \cdot \vec{\sigma}^{(1)}}{2} \frac{1 - \vec{n} \cdot \vec{\sigma}^{(2)}}{2} \right\rangle ,$$

$$\text{prob}(\uparrow\downarrow) = \left\langle \frac{1 + \vec{e} \cdot \vec{\sigma}^{(1)}}{2} \frac{1 - \vec{n} \cdot \vec{\sigma}^{(2)}}{2} \right\rangle ,$$

$$\text{prob}(\downarrow\uparrow) = \left\langle \frac{1 - \vec{e} \cdot \vec{\sigma}^{(1)}}{2} \frac{1 + \vec{n} \cdot \vec{\sigma}^{(2)}}{2} \right\rangle , \qquad (2.18.3)$$

where, for instance $\text{prob}(\uparrow\downarrow)$ is the probability that Tom's atom is found as $\uparrow$ and Jerry's as $\downarrow$. Accordingly,

$$\text{prob}(\uparrow\uparrow \text{ or } \downarrow\downarrow) = \text{prob}(\uparrow\uparrow) + \text{prob}(\downarrow\downarrow)$$
$$= \left\langle \frac{1}{2}\left(1 + \vec{e} \cdot \vec{\sigma}^{(1)}\, \vec{n} \cdot \vec{\sigma}^{(2)}\right) \right\rangle \qquad (2.18.4)$$

and

$$\text{prob}(\uparrow\downarrow \text{ or } \downarrow\uparrow) = \text{prob}(\uparrow\downarrow) + \text{prob}(\downarrow\uparrow)$$
$$= \left\langle \frac{1}{2}\left(1 - \vec{e} \cdot \vec{\sigma}^{(1)}\, \vec{n} \cdot \vec{\sigma}^{(2)}\right) \right\rangle , \qquad (2.18.5)$$

so that

$$C(\vec{e}, \vec{n}) = \left\langle \vec{e} \cdot \vec{\sigma}^{(1)}\, \vec{n} \cdot \vec{\sigma}^{(2)} \right\rangle = \text{tr}\left\{ \vec{e} \cdot \vec{\sigma}^{(1)}\, \vec{n} \cdot \vec{\sigma}^{(2)} \rho \right\} , \qquad (2.18.6)$$

where ρ is the statistical operator that specifies the state of the atom pairs emitted by the source.

The source could, for instance, emit the atoms in a state with $|\uparrow_z \downarrow_z\rangle$, that is: Tom's atom $\uparrow_z$ and Jerry's atom $\downarrow_z$. Or in a state with ket $|\downarrow_z \uparrow_z\rangle$, now Tom's atom is $\downarrow_z$ and Jerry's atom is $\uparrow_z$. Of course, there are also the possibilities of $|\uparrow_z \uparrow_z\rangle$, or $|\downarrow_z \downarrow_z\rangle$, or any other ket comprised of them as a weighted sum,

$$|\ \rangle = |\uparrow_z \uparrow_z\rangle \alpha + |\downarrow_z \downarrow_z\rangle \beta + |\uparrow_z \downarrow_z\rangle \gamma + |\downarrow_z \uparrow_z\rangle \delta , \qquad (2.18.7)$$

where the probability amplitudes $\alpha, \ldots, \delta$ are normalized in accordance with

$$|\alpha|^2 + |\beta|^2 + |\gamma|^2 + |\delta|^2 = 1. \qquad (2.18.8)$$

These are simple self-suggesting extensions of the formalism that we have developed for single atoms. The main difference is that we now have four basic options, two for each atom.

We are aiming at establishing a situation in which the Bell inequality (1.2.11) is violated by quantum correlations that are substantially stronger than any classical correlation could ever be. For this purpose it is simplest to consider a source that emits the two-atom state with ket

$$| \, \rangle = \frac{1}{\sqrt{2}} \Big(|\uparrow_z \downarrow_z \rangle - |\downarrow_z \uparrow_z \rangle \Big). \tag{2.18.9}$$

2-40   Express this in terms of up/down states for the x-direction, that is as a sum of $|\uparrow_x \uparrow_x \rangle, |\uparrow_x \downarrow_x \rangle, \cdots$.

In a state of this kind, the atoms have no individual properties. Tom's atom can either be $\uparrow_z$ or $\downarrow_z$, and the same applies to Jerry's atom. But jointly, the two atoms have the distinct objective property that if one is $\uparrow_z$, the other will be $\downarrow_z$. Which one is $\uparrow_z$, and which is $\downarrow_z$, is not predictable, but that they are paired in this fashion is. Put differently, if Tom chooses $\vec{e} \mathrel{\hat{=}} \begin{pmatrix} 0 \\ 0 \\ 1 \end{pmatrix}$ and Jerry chooses $\vec{n} \mathrel{\hat{=}} \begin{pmatrix} 0 \\ 0 \\ 1 \end{pmatrix}$, the measurement results will either be $+1$ and -1 or -1 and $+1$, but never $+1$ and $+1$ or -1 and -1. For this setting, then, we have

$$C(\vec{e}, \vec{n}) = -1 \quad \text{if} \quad \vec{e} = \vec{n} \mathrel{\hat{=}} \begin{pmatrix} 0 \\ 0 \\ 1 \end{pmatrix}. \tag{2.18.10}$$

The statistical operator $\rho = | \, \rangle\langle \, |$ for this two-atom state is

$$\begin{aligned} \rho &= \frac{1}{2} \Big(|\uparrow_z \downarrow_z \rangle - |\downarrow_z \uparrow_z \rangle \Big) \Big(\langle \uparrow_z \downarrow_z | - \langle \downarrow_z \uparrow_z | \Big) \\ &= \frac{1}{2} |\uparrow_z \downarrow_z \rangle \langle \uparrow_z \downarrow_z | + \frac{1}{2} |\downarrow_z \uparrow_z \rangle \langle \downarrow_z \uparrow_z | \\ &\quad - \frac{1}{2} |\uparrow_z \downarrow_z \rangle \langle \downarrow_z \uparrow_z | - \frac{1}{2} |\downarrow_z \uparrow_z \rangle \langle \uparrow_z \downarrow_z |, \end{aligned} \tag{2.18.11}$$

which we express in terms of $\vec{\sigma}^{(1)}$ and $\vec{\sigma}^{(2)}$ by first noting that, for each atom by itself,

$$|\uparrow_z\rangle\langle\uparrow_z| = \frac{1}{2}(1 + \sigma_z),$$

$$|\downarrow_z\rangle\langle\downarrow_z| = \frac{1}{2}(1 - \sigma_z),$$

$$|\uparrow_z\rangle\langle\downarrow_z| = \frac{1}{2}(\sigma_x + i\sigma_y),$$

$$|\downarrow_z\rangle\langle\uparrow_z| = \frac{1}{2}(\sigma_x - i\sigma_y).\qquad(2.18.12)$$

Thus

$$|\uparrow_z\uparrow_z\rangle\langle\uparrow_z\uparrow_z| = \frac{1}{2}\left(1 + \sigma_z^{(1)}\right)\frac{1}{2}\left(1 + \sigma_z^{(2)}\right),\qquad(2.18.13)$$

for instance, and we get

$$
\begin{aligned}
\rho &= \frac{1}{2}\frac{1+\sigma_z^{(1)}}{2}\frac{1-\sigma_z^{(2)}}{2} + \frac{1}{2}\frac{1-\sigma_z^{(1)}}{2}\frac{1+\sigma_z^{(2)}}{2}\\
&\quad - \frac{1}{2}\frac{\sigma_x^{(1)}+i\sigma_y^{(1)}}{2}\frac{\sigma_x^{(2)}-i\sigma_y^{(2)}}{2} - \frac{1}{2}\frac{\sigma_x^{(1)}-i\sigma_y^{(1)}}{2}\frac{\sigma_x^{(2)}+i\sigma_y^{(2)}}{2}\\
&= \frac{1}{4}\left(1 - \sigma_x^{(1)}\sigma_x^{(2)} - \sigma_y^{(1)}\sigma_y^{(2)} - \sigma_z^{(1)}\sigma_z^{(2)}\right)\\
&= \frac{1}{4}\left(1 - \vec{\sigma}^{(1)}\cdot\vec{\sigma}^{(2)}\right).\qquad(2.18.14)
\end{aligned}
$$

Note two things about this. First, the identity is here multiplied by $\frac{1}{4}$ because we now have a four-dimensional ket space (and bra space) so that the matrix representation is by 4×4 matrices and $\text{tr}\{1\} = 4$. Second, there is no difference between x, y, and z in the compact final form of ρ, which implies immediately that we also have

$$C(\vec{e},\vec{n}) = -1 \quad \text{if} \quad \vec{e} = \vec{n} \cong \begin{pmatrix} 1 \\ 0 \\ 0 \end{pmatrix} \quad \text{or} \quad \vec{e} = \vec{n} \cong \begin{pmatrix} 0 \\ 1 \\ 0 \end{pmatrix}$$

$$\text{or} \quad \vec{e} = \vec{n} = \text{any direction}.\qquad(2.18.15)$$

Despite the original reference to the z direction in (2.18.9), in fact no spatial direction is singled out by this two-atom state.

Now, the trace in

$$C(\vec{e},\vec{n}) = \text{tr}\left\{\vec{e}\cdot\vec{\sigma}^{(1)}\,\vec{n}\cdot\vec{\sigma}^{(2)}\frac{1}{4}\left(1 - \vec{\sigma}^{(1)}\cdot\vec{\sigma}^{(2)}\right)\right\}\qquad(2.18.16)$$

refers to both atoms, $\mathrm{tr}\{\ \} = \mathrm{tr}_{1,2}\{\ \} = \mathrm{tr}_1\{\mathrm{tr}_2\{\ \}\}$, and we evaluate it successively

$$C(\vec{e}, \vec{n}) = \mathrm{tr}_1\left\{ \vec{e} \cdot \vec{\sigma}^{(1)} \mathrm{tr}_2\left\{ \vec{n} \cdot \vec{\sigma}^{(2)} \frac{1}{4}\left(1 - \vec{\sigma}^{(1)} \cdot \vec{\sigma}^{(2)}\right) \right\} \right\}. \qquad (2.18.17)$$

Recalling that, for a single atom,

$$\mathrm{tr}\{\vec{a} \cdot \vec{\sigma}\} = 0, \qquad \mathrm{tr}\left\{ \vec{a} \cdot \vec{\sigma}\ \vec{b} \cdot \vec{\sigma} \right\} = 2\vec{a} \cdot \vec{b}, \qquad (2.18.18)$$

and noting that $\vec{\sigma}^{(1)}$ is just like a numerical vector for atom 2, we have

$$\mathrm{tr}_2\left\{ \vec{n} \cdot \vec{\sigma}^{(2)} \frac{1}{4}\left(1 - \vec{\sigma}^{(1)} \cdot \vec{\sigma}^{(2)}\right) \right\} = -\frac{1}{2}\vec{n} \cdot \vec{\sigma}^{(1)} \qquad (2.18.19)$$

and then

$$C(\vec{e}, \vec{n}) = \mathrm{tr}_1\left\{ \vec{e} \cdot \vec{\sigma}^{(1)} \left(-\frac{1}{2}\vec{n} \cdot \vec{\sigma}^{(1)}\right) \right\} = -\vec{e} \cdot \vec{n}, \qquad (2.18.20)$$

fully consistent with the observation in (2.18.15) that $C(\vec{e}, \vec{n}) = -1$ for $\vec{e} = \vec{n} = $ any direction.

The left-hand side of the Bell inequality (1.2.11), now with $a \to \vec{e}_1$, $a' \to \vec{e}_2$ and $b \to \vec{n}_1$, $b' \to \vec{n}_2$ as fits the present situation in which parameter settings are choices of directions, is then given by

$$L \equiv |C(\vec{e}_1, \vec{n}_1) - C(\vec{e}_1, \vec{n}_2)| + |C(\vec{e}_2, \vec{n}_1) + C(\vec{e}_2, \vec{n}_2)|$$
$$= |\vec{e}_1 \cdot (\vec{n}_1 - \vec{n}_2)| + |\vec{e}_2 \cdot (\vec{n}_1 + \vec{n}_2)|, \qquad (2.18.21)$$

and we wish to choose the four directions such that we maximize the value of L. We do this in two steps. First, for given $\vec{n}_1$ and $\vec{n}_2$, we optimize $\vec{e}_1$ and $\vec{e}_2$ clearly by having

$$\vec{e}_1 = \frac{\vec{n}_1 - \vec{n}_2}{|\vec{n}_1 - \vec{n}_2|} \qquad \text{and} \qquad \vec{e}_2 = \frac{\vec{n}_1 + \vec{n}_2}{|\vec{n}_1 + \vec{n}_2|}, \qquad (2.18.22)$$

so that

$$\max L = |\vec{n}_1 - \vec{n}_2| + |\vec{n}_1 + \vec{n}_2|$$
$$= \sqrt{2 - 2\vec{n}_1 \cdot \vec{n}_2} + \sqrt{2 + 2\vec{n}_1 \cdot \vec{n}_2}, \qquad (2.18.23)$$

or with $\vec{n}_1 \cdot \vec{n}_2 = \cos\theta$,

$$\begin{aligned}
\max L &= 2\sin\frac{\theta}{2} + 2\cos\frac{\theta}{2} \\
&= 2\sqrt{2}\Big(\underbrace{\frac{1}{\sqrt{2}}}_{=\cos\frac{\pi}{4}}\cos\frac{\theta}{2} + \underbrace{\frac{1}{\sqrt{2}}}_{=\sin\frac{\pi}{4}}\sin\frac{\theta}{2} \Big) \\
&= 2\sqrt{2}\cos\left(\frac{\pi}{4} - \frac{\theta}{2}\right).
\end{aligned}$$
(2.18.24)

The absolutely largest value is obtained for $\theta = \pi/2$ when

$$\max L = 2\sqrt{2},$$
(2.18.25)

clearly exceeding the upper bound of the Bell inequality (1.2.11), $L \leq 2$. We conclude, that the Bell inequality does not apply to quantum correlations of the type built into two-atom states of the kind exemplified by (2.18.9) where the individual atoms have no properties of their own, whereas the pair of atoms has distinct statistical properties. Such an intimate link is called *entanglement*. One says the magnetic degrees of freedom of the two atoms are entangled or, more sloppily, that the atoms are entangled.

This closes the argument given in Section 1.2, and ultimately justifies the conclusion on page 8 about the nonexistence of any mechanism that would decide the outcome of a quantum measurement. The violation of Bell's inequality is an established experimental fact. It has become almost a routine experiment to check that $L \leq 2\sqrt{2}$ as part of the calibration procedure in experiments of a certain type.

2.19   State reduction

Another point worth mentioning is the following. Suppose Tom is unaware of Jerry and his measurements and just sees the atoms emitted by the source to the left. They are a mixture of $\uparrow_z$ and $\downarrow_z$, or of $\uparrow_x$ and $\downarrow_x$, or $\ldots$, to him,

$$\begin{aligned}
\rho_1^{(\mathrm{T})} &= \frac{1}{2}\frac{1+\sigma_z^{(1)}}{2} + \frac{1}{2}\frac{1-\sigma_z^{(1)}}{2} = \frac{1}{2} \\
&= \frac{1}{2}\frac{1+\sigma_x^{(1)}}{2} + \frac{1}{2}\frac{1-\sigma_x^{(1)}}{2} = \frac{1}{2},
\end{aligned}$$
(2.19.1)

and Tom makes statistical predictions about the next atom consistent with this statistical operator. How about Jerry, then? That depends. As long as he has not measured his atom on the right, Jerry also uses

$$\rho_1^{(J)} = \frac{1}{2} \tag{2.19.2}$$

to predict the statistics of measurement results for Tom's atom. But if he measures his atom along z and finds $\uparrow_z$, say, then Jerry uses

$$\rho_1^{(J)} = \frac{1 - \sigma_z^{(1)}}{2} \tag{2.19.3}$$

since he knows that Tom's atom is $\downarrow_z$ under these circumstances. And if he measures his atom along x and finds $\downarrow_x$, say, then Jerry uses

$$\rho_1^{(J)} = \frac{1 + \sigma_x^{(1)}}{2} \tag{2.19.4}$$

because he knows that Tom's atom has to be $\uparrow_x$ under these circumstances. The transitions

$$\rho_1^{(J)} = \frac{1}{2} \rightarrow \begin{cases} \rho_1^{(J)} = \dfrac{1 - \sigma_z^{(1)}}{2} \\ \rho_1^{(J)} = \dfrac{1 + \sigma_x^{(1)}}{2} \end{cases} \tag{2.19.5}$$

are examples of *state reduction*, a technical procedure by which additional, newly acquired knowledge is taken into account in order to adjust our statistical predictions. State reduction is not a physical process, nothing happens to Tom's atom when Jerry measures his, it is a mental process reflecting that Jerry has identified Tom's atom as a member of a particular subensemble.

It is thus possible that Tom uses

$$\rho_1^{(T)} = \frac{1}{2} \tag{2.19.6}$$

and Jerry uses

$$\rho_1^{(J)} = \frac{1 - \sigma_z^{(1)}}{2} \tag{2.19.7}$$

for statistical prediction about one and the same next atom, and both make statistically correct predictions. This teaches us the lesson that the statistical operator is not a property of the atom, such as its mass or its electric charge, but a theoretical tool of the physicist who talks about the atom professionally, taking into account all *he* knows about the atom. Another physicist may have different knowledge, and then she will use another statistical operator, one that correctly reflects *her* knowledge.

2.20   Measurements with more than two outcomes

For a single Ag atom there are two outcomes in a Stern–Gerlach measurement, for a pair of atoms there are $2^2 = 4$ outcomes altogether for measurements on both atoms, and so forth: $2^3 = 8$ for 3 atoms, $2^4 = 16$ for 4 atoms, ... We then need to extend the formalism such that there are $4, 8, 16, \ldots$ pairwise orthogonal ket vectors, rather than just 2. Further we note that instead of the magnetic properties of the atom, we could be interested in other physical phenomena, for which there are more than two choices to begin with. Think, for example, of the total electric charge of a collection of atoms. That could be $0, \pm 1, \pm 2, \pm 3, \ldots$ units of electric charge, with no *a priori* restriction on these integers — an example of a physical property that would require an infinite set of pairwise orthogonal kets in the mathematical formalism. And then there is the possibility of a continuum of measurement result, as one gets it when determining the position of an object, for example. Here, too, we need an infinite number of pairwise orthogonal kets, but a continuous set, rather than a discrete one as in the previous example of electric charge.

Further, in those situations where a finite number of kets suffice, it need not always be a power of 2, one can also have physical properties that have $3, 5, 6, \ldots$ possible values as measurement results. For instance, there are atoms (orthohelium is an example), such that a Stern–Gerlach apparatus splits a beam of them in three.

It is clear, then, that we must generalize the formalism if we want, as we do, to deal with these other physical properties. For the time being, we shall assume that the number of possible measurement results is a finite number n, and shall address the case of infinitely many measurements results later, in Chapter 4. So the symbolic generalization of a Stern–Gerlach apparatus is of the sort

We assign numbers $a_1, \ldots, a_n$ as measurement results to the various outcomes, and have kets

$$|a_1\rangle, |a_2\rangle, \ldots, |a_n\rangle \tag{2.20.1}$$

associated with them. These exclude each other, just as $\uparrow_z$ and $\downarrow_z$ did in (2.8.4), which is expressed by their orthogonality

$$\langle a_j | a_k \rangle = \delta_{jk} = \begin{cases} 1 & \text{if } j = k, \\ 0 & \text{if } j \neq k, \end{cases} \tag{2.20.2}$$

for $j, k = 1, 2, \ldots, n$. Since this also contains the statement of normalization to unit length, $\langle a_j | a_j \rangle = 1$ for $j = 1, 2, \ldots, n$, the generalization of (2.8.3), we have an *orthonormality relation*. The symbol δ_{jk} is known as the *Kronecker delta symbol* and is named after Leopold Kronecker.

We did not overlook any possibilities, which is to say that the kets $|a_j\rangle$ are complete, as expressed by the *completeness relation*

$$\sum_{j=1}^{n} |a_j\rangle\langle a_j| = 1. \tag{2.20.3}$$

Applied to the arbitrary ket $| \ \rangle$,

$$| \ \rangle = \sum_{j=1}^{n} |a_j\rangle\langle a_j| \ \rangle, \tag{2.20.4}$$

it states that $| \ \rangle$ is a weighted sum of the basic kets, with the probability amplitudes $\langle a_j | \ \rangle$ as weights. Their squares $\left|\langle a_j | \ \rangle\right|^2$ are then the probabilities that the measurement has outcome a_j if the ket $| \ \rangle$ correctly describes the incoming atoms. These probabilities sum up to unity,

$$\sum_{j} \left|\langle a_j | \ \rangle\right|^2 = \sum_{j} \langle \ |a_j\rangle\langle a_j| \ \rangle$$
$$= \langle \ |\left(\sum_{j=1}^{n} |a_j\rangle\langle a_j|\right)| \ \rangle$$
$$= \langle \ |1| \ \rangle = \langle \ | \ \rangle = 1, \tag{2.20.5}$$

which is just another way of saying that no possibility was left out.

Having assigned numbers a_j to the outcomes, we naturally use them as the eigenvalues of the operator A that we associate with the measured property,

$$A|a_j\rangle = |a_j\rangle a_j,\tag{2.20.6}$$

and the kets $|a_j\rangle$ are the respective eigenkets. The bras $\langle a_j|$ are the eigen-bras, quite analogously,

$$\langle a_j|A = a_j\langle a_j|.\tag{2.20.7}$$

Note, however, that these two eigenvector equations are not adjoints of each other. Rather the adjoint statement of $A|a_j\rangle = |a_j\rangle a_j$ is

$$\langle a_j|A^\dagger = a_j^*\langle a_j|\tag{2.20.8}$$

consistent with

$$A^\dagger = \left(\sum_{j=1}^n |a_j\rangle a_j\langle a_j|\right)^\dagger = \sum_{j=1}^n |a_j\rangle a_j^*\langle a_j|\tag{2.20.9}$$

which is an immediate consequence of

$$\left(|1\rangle\lambda\langle 2|\right)^\dagger = |2\rangle\lambda^*\langle 1|,\tag{2.20.10}$$

a simple generalization of (2.8.15) that is used here for $|1\rangle = |a_j\rangle$, $\lambda = a_j$, $\langle 2| = \langle a_j|$, and $j = 1,\ldots,n$. In other words, the kets $|a_j\rangle$ are also eigenkets of $A^\dagger$ with eigenvalues a_j^*, and the bras $\langle a_j|$ are eigenbras of $A^\dagger$ with these eigenvalues.

2-41 What are the eigenvalues, eigenkets, eigenbras of $A^\dagger A$ and $AA^\dagger$? Show that $A^\dagger A = AA^\dagger$. Such operators are called *normal*; they represent, quite generally, physical properties.

Rather than the numbers a_j, we could have assigned another set of numbers to the n different possibilities, such as $f(a_j)$, where the function f must be well defined for all arguments $a_1, a_2, \ldots, a_n$ but we need not specify $f(x)$ for other values of x. Then, instead of A given above, we would use the operator

$$\sum_{j=1}^n |a_j\rangle f(a_j)\langle a_j|\tag{2.20.11}$$

as the mathematical symbol for the physical property measured by the apparatus. It is systematic to regard the right-hand side as defining $f(A)$, the corresponding function of operator A,

$$f(A) = f\left(\sum_{j=1}^{n}|a_j\rangle a_j\langle a_j|\right) = \sum_{j=1}^{n}|a_j\rangle f(a_j)\langle a_j|. \qquad (2.20.12)$$

The notation is suggestive, and as such it must not be misleading, that is to say that whenever we can give $f(A)$ meaning in a different way, the two definitions must coincide. So let us look at A^2,

$$A^2 = \left(\sum_{j=1}^{n}|a_j\rangle a_j\langle a_j|\right)^2$$

$$= \sum_{j=1}^{n}\sum_{k=1}^{n}|a_j\rangle a_j\underbrace{\langle a_j|a_k\rangle}_{=\delta_{jk}} a_k\langle a_k|$$

$$= \sum_{j=1}^{n}|a_j\rangle a_j^2\langle a_j|, \qquad (2.20.13)$$

and by induction we conclude that

$$A^N = \left(\sum_{j=1}^{n}|a_j\rangle a_j\langle a_j|\right)^N = \sum_{j=1}^{n}|a_j\rangle a_j^N\langle a_j|. \qquad (2.20.14)$$

Accordingly,

$$f(A) = \sum_{j}|a_j\rangle f(a_j)\langle a_j| \qquad (2.20.15)$$

is consistent if $f(A)$ is a power of A, and then it is also consistent when $f(A)$ is a polynomial of A. Since almost all functions of potential interest to us can be approximated by, or related to, polynomials, we accept (2.20.15) as the generally valid definition of a function of A. The right-hand side of (2.20.15) is called the *spectral decomposition* of $f(A)$, and we see the spectral decompositions of A itself and of $A^\dagger$ in (2.20.9). Thereby we keep in mind that $f(x)$ must be well defined for $x = a_1$, $x = a_2$, ..., $x = a_n$, while other values of x are irrelevant. As a consequence of (2.20.15), two functions of A, $f_1(A)$ and $f_2(A)$, say, are equal if $f_1(a) = f_2(a)$ for all eigenvalues a of A, and only in that situation.

2-42 Express $f_1(\sigma_x) = e^{i\varphi\sigma_x}$ and $f_2(\sigma_x) = \dfrac{1 + i\tau\sigma_x}{1 - i\tau\sigma_x}$ with real parameters φ and τ, as linear functions of σ_x. How must φ and τ be related to each other in order to ensure $f_1(\sigma_x) = f_2(\sigma_x)$?

A second physical property, B, has measurement results $b_1, b_2, \ldots, b_n$, eigenbras $\langle b_j |$, eigenkets $| b_k \rangle$, etc. and for all relations stated above about A, there are, of course, the analogous statements about B:

$$\text{eigenkets: } B | b_k \rangle = | b_k \rangle b_k \,,$$

$$\text{eigenbras: } \langle b_k | B = b_k \langle b_k | \,,$$

$$\text{operator } B: \ B = \sum_{k=1}^{n} | b_k \rangle b_k \langle b_k | \,,$$

$$\text{function of } B: \ g(B) = \sum_{k=1}^{n} | b_k \rangle g(b_k) \langle b_k | \,,$$

$$\text{orthonormality: } \langle b_j | b_k \rangle = \delta_{jk} \,,$$

$$\text{completeness: } \sum_{k=1}^{n} | b_k \rangle \langle b_k | = 1 \,. \tag{2.20.16}$$

And, in addition, we have statements involving both sets of kets and bras, in particular

$$\langle a_j | b_k \rangle = \langle b_k | a_j \rangle^* \,, \tag{2.20.17}$$

the statement of Exercise 2-11 on page 40 that reversed probability amplitudes are complex conjugates of each other.

The fundamental symmetry of the probabilities

$$\begin{aligned} \text{prob}(b_k \leftarrow a_j) &= \left| \langle b_k | a_j \rangle \right|^2 \\ &= \left| \langle a_j | b_k \rangle \right|^2 = \text{prob}(a_j \leftarrow b_k) \end{aligned} \tag{2.20.18}$$

expresses an important fact about quantum processes. In the pictorial symbolization that generalizes the original Stern–Gerlach discussion, we would draw

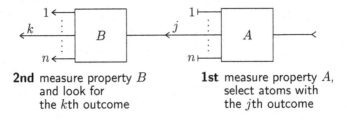

2nd measure property B
and look for
the kth outcome

1st measure property A,
select atoms with
the jth outcome

for the experiment in which we measure $\text{prob}(b_k \leftarrow a_j)$, and

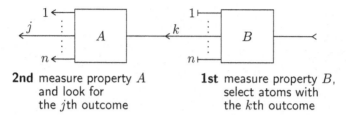

2nd measure property A and look for the jth outcome   **1st** measure property B, select atoms with the kth outcome

for the experiment in which we measure $\text{prob}(a_j \leftarrow b_k)$.

These are two quite different experiments, if properties A and B are quite different aspects of an atomic system, such as, say, its energy and its angular momentum. Nevertheless, and irrespective of which two properties we are dealing with, the equality of the probabilities (2.20.18) is always true. It seems, however, that this fundamental prediction of the quantum-mechanical formalism has never been studied systematically in an experiment, and so we must rely on circumstantial evidence provided by the overwhelmingly successful performance of quantum mechanics in much more complicated situations, where this symmetry is an essential ingredient in the calculations.

2.21 Unitary operators

Now, having the A states and the B states at hand, we can ask for the operator that relates them to each other: What is U_{ba} in

$$U_{ba}|a_k\rangle = |b_k\rangle \qquad \text{for} \quad k = 1, 2, \ldots, n\,? \tag{2.21.1}$$

Multiply by $\langle a_k|$ from the right, sum over k, and exploit the completeness relation for the A states,

$$\sum_k U_{ba}|a_k\rangle\langle a_k| = U_{ba}\underbrace{\sum_k |a_k\rangle\langle a_k|}_{=\,1}$$

$$= \sum_k |b_k\rangle\langle a_k|\,, \tag{2.21.2}$$

to arrive at

$$U_{ba} = \sum_{k=1}^{n} |b_k\rangle\langle a_k|\,. \tag{2.21.3}$$

It follows that

$$\langle b_k | U_{ba} = \langle a_k | \qquad \text{for} \quad k = 1, 2, \ldots, n \qquad (2.21.4)$$

holds as well.

2-43 If properties A and B are physically similar in the sense that $a_j = b_j$ for $j = 1, 2, \ldots, n$, then show that

$$B U_{ba} = U_{ba} A$$

holds.

2-44 Show that

$$U_{ba} f(A) = f(B) U_{ba}$$

under these circumstances.

The reverse mapping

$$U_{ab} | b_k \rangle = | a_k \rangle \qquad (2.21.5)$$

is accomplished by the operator

$$U_{ab} = \sum_{k=1}^{n} | a_k \rangle \langle b_k | \qquad (2.21.6)$$

which is both the inverse of U_{ba},

$$U_{ba} U_{ab} = U_{ab} U_{ba} = 1 : \quad U_{ab} = U_{ba}^{-1}, \qquad (2.21.7)$$

and the adjoint of U_{ba},

$$U_{ab} = \sum_k | a_k \rangle \langle b_k | = \sum_k \left(| b_k \rangle \langle a_k | \right)^{\dagger}$$
$$= \left(\sum_k | b_k \rangle \langle a_k | \right)^{\dagger} = U_{ba}^{\dagger}. \qquad (2.21.8)$$

It follows that U_{ba} has the characteristic property that its adjoint is its inverse,

$$U_{ba} (U_{ba})^{\dagger} = 1, \qquad (U_{ba})^{\dagger} U_{ba} = 1, \qquad (2.21.9)$$

and the same is, of course, also true for U_{ab},

$$U_{ab} (U_{ab})^{\dagger} = 1, \qquad (U_{ab})^{\dagger} U_{ab} = 1. \qquad (2.21.10)$$

Operators of this kind are called *unitary*. They are the generalization of rotation operators for 3-vectors in ordinary space. For, a rotation leaves all scalar properties unchanged,

$$
\left.\begin{array}{c} \vec{r} \xrightarrow{\text{rotation}} \vec{r}\,' \\[2mm] \vec{s} \xrightarrow{\text{rotation}} \vec{s}\,' \end{array}\right\} \qquad \vec{r} \cdot \vec{s} = \vec{r}\,' \cdot \vec{s}\,' \tag{2.21.11}
$$

for any pair of 3-vectors. The corresponding statement about unitary operators is the invariance of brackets,

$$
\left.\begin{array}{c} |1\rangle \xrightarrow{\text{unitary}} |1'\rangle \\[2mm] |2\rangle \xrightarrow{\text{unitary}} |2'\rangle \end{array}\right\} \qquad \langle 1|2\rangle = \langle 1'|2'\rangle \tag{2.21.12}
$$

for all kets $|1\rangle, |2\rangle$.

Indeed, if we have

$$
\begin{aligned} |1\rangle &\longrightarrow |1'\rangle = U|1\rangle, \\ |2\rangle &\longrightarrow |2'\rangle = U|2\rangle, \end{aligned} \tag{2.21.13}
$$

then

$$
\begin{aligned} \langle 1|2\rangle \longrightarrow \langle 1'|2'\rangle &= \left(\langle 1|U^\dagger\right)\left(U|2\rangle\right) \\ &= \langle 2|(U^\dagger U)|2\rangle \\ &= \langle 1|2\rangle \end{aligned} \tag{2.21.14}
$$

holds for *any* choice of $|1\rangle$ and $|2\rangle$, if

$$
U^\dagger U = 1, \tag{2.21.15}
$$

that is: if U is unitary. In particular it follows that a unitary operator maps any given set of orthonormal kets onto another set of this sort. The unitary operators U_{ba} and U_{ab} from above illustrate this, inasmuch as they were constructed such that they map the set of $|a_k\rangle$s and the set of $|b_k\rangle$s onto each other.

2-45 As an immediate extension of Exercise 2-44, show that

$$
U^{-1} f(A) U = f(U^{-1} A U)
$$

for any operator A representing a physical property and any unitary operator U.

Given some unitary operator, we denote its eigenvalues by u_k, the corresponding eigenkets by $|u_k\rangle$, and the eigenbras by $\langle u_k|$, so that

$$U|u_k\rangle = |u_k\rangle u_k \,,$$
$$\langle u_k|U = u_k\langle u_k| \,, \tag{2.21.16}$$

which we supplement by the adjoint statements,

$$\langle u_k|U^\dagger = u_k^*\langle u_k| \,,$$
$$U^\dagger|u_k\rangle = |u_k\rangle u_k^* \,. \tag{2.21.17}$$

We combine them in

$$u_k^*\langle u_k|u_k\rangle u_k = \langle u_k|U^\dagger U|u_k\rangle$$
$$= \langle u_k|u_k\rangle \tag{2.21.18}$$

to conclude that

$$u_k^* u_k = 1 \qquad \text{or} \qquad |u_k| = 1 \qquad \text{for all} \quad k = 1,\ldots,n\,. \tag{2.21.19}$$

That is: all eigenvalues of a unitary operator are complex numbers with unit modulus, they are phase factors,

$$u_k = e^{i\varphi_k} \qquad \text{with a real phase } \varphi_k. \tag{2.21.20}$$

2-46 The converse is also true: Show that

$$f(A) = \sum_{k=1}^n |a_k\rangle f(a_k)\langle a_k| \tag{2.21.21}$$

is unitary if $|f(a_k)|^2 = 1$ for $k = 1, 2, \ldots, n$.

2.22   Hermitian operators

Given U, and thus its eigenvalues u_k, the phases φ_k are not determined uniquely, there is the obvious arbitrariness of adding any integer multiple of 2π to φ_k without changing the value of u_k. But, once we have chosen a particular set of φ_k, we can construct the new operator

$$\Phi = \sum_{k=1}^n |u_k\rangle \varphi_k\langle u_k| \,, \tag{2.22.1}$$

which is such that

$$U = e^{i\Phi}. \tag{2.22.2}$$

The fact that the phases φ_k are real — they are the eigenvalues of Φ, of course — implies that Φ is equal to its adjoint,

$$\begin{aligned}
\Phi^\dagger &= \sum_{k=1}^{n} \Big(|u_k\rangle \varphi_k \langle u_k| \Big)^\dagger \\
&= \sum_{k=1}^{n} |u_k\rangle \varphi_k^* \langle u_k| = \sum_{k=1}^{n} |u_k\rangle \varphi_k \langle u_k| \\
&= \Phi.
\end{aligned} \tag{2.22.3}$$

It is thus an example of another important class of operators; those that are adjoints of themselves. Such operators are called *selfadjoint* operators, or *hermitian* operators, in honor of Charles Hermite.

The very important relation

$$U = e^{iH} \qquad \text{with} \quad H = H^\dagger \quad \text{(hermitian)}$$
$$\text{and} \quad U = (U^\dagger)^{-1} \quad \text{(unitary)} \tag{2.22.4}$$

goes both ways: If U is unitary, then H is hermitian (see above), and if H is hermitian, then U is unitary. To demonstrate the latter, let us first note the eigenvalue equations for H,

$$\begin{aligned}
H|h_k\rangle &= |h_k\rangle h_k, \\
\langle h_k|H &= h_k \langle h_k|,
\end{aligned} \tag{2.22.5}$$

where the possible h_ks are the eigenvalues and $|h_k\rangle$, $\langle h_k|$ the corresponding eigenkets and eigenbras. Their adjoint statements

$$\begin{aligned}
h_k^* \langle h_k| &= \langle h_k|H^\dagger = \langle h_k|H, \\
|h_k\rangle h_k^* &= H^\dagger|h_k\rangle = H|h_k\rangle,
\end{aligned} \tag{2.22.6}$$

imply the reality of the eigenvalues, $h_k = h_k^*$. Then,

$$\begin{aligned}
e^{iH} &= \exp\left(i \sum_{k=1}^{n} |h_k\rangle h_k \langle h_k| \right) \\
&= \sum_{k=1}^{n} |h_k\rangle e^{ih_k} \langle h_k|
\end{aligned} \tag{2.22.7}$$

is unitary, because we have the situation of Exercise 2-46 with $A = H$ and $f(A) = \mathrm{e}^{\mathrm{i}A}$.

2-47 An operator is known to be both unitary and hermitian. What can you say about its eigenvalues? Show that

$$A = \mathrm{e}^{\mathrm{i}\frac{\pi}{2}(1 - A)}$$

for all such hermitian-unitary operators A.

Chapter 3

Dynamics:
How Quantum Systems Evolve

3.1 Schrödinger equation

Compare the following two situations that harken back to the end of Section 2.4. In the first scenario atoms arrive "+ in z" and σ_z is measured, with outcome +1 for all of them:

$$
\begin{array}{l}
\text{all } +1 \\
\text{none } -1
\end{array}
\boxed{\begin{array}{c}\text{measure}\\ \sigma_z\end{array}}
\longleftarrow \uparrow_z
$$

or with outcome −1 for all if they arrive as $\downarrow_z$:

$$
\begin{array}{l}
\text{none } +1 \\
\text{all } -1
\end{array}
\boxed{\begin{array}{c}\text{measure}\\ \sigma_z\end{array}}
\longleftarrow \downarrow_z
$$

In the second scenario, a homogeneous magnetic field of appropriate strength and direction (along the y axis) effects

$$\uparrow_z \longrightarrow \uparrow_x \quad \text{and} \quad \downarrow_z \longrightarrow \downarrow_x \tag{3.1.1}$$

and then σ_x is measured, so that

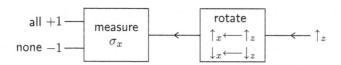

or

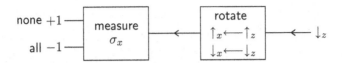

The measurement *result* is unchanged, you get predictably $+1$ for $\uparrow_z$ arriving and predictably -1 for $\downarrow_z$ arriving, although the apparatus is quite different. We can summarize these observations by stating that

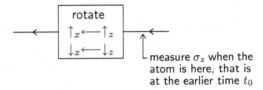

and

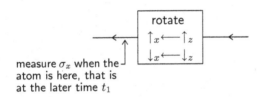

are equivalent measurements, they give the same results. Now recognizing that the instant at which you perform the measurement is crucial, we indicate the time dependence as an argument, such that $\sigma_z(t)$, for example, refers to measuring in the z direction at time t. In the above situation, then, we have

$$\sigma_x(t_1) = \sigma_z(t_0)\,. \tag{3.1.2}$$

This describes the overall effect, the "before to after" change, resulting from the action of the magnetic field, but clearly there must be intermediate stages with an altogether smooth transition from t_0 to t_1.

To formulate the equations of motion that describe infinitesimal changes during infinitesimal time intervals, we consider the general case where operator A has n different eigenvalues $a_1, \ldots, a_n$ and corresponding eigenkets and eigenbras. To begin with we stick to the case illustrated by the $\sigma_z(t_0) = \sigma_x(t_1)$ example above, where the possible measurement results

($+1$ or -1) do not change in time. Here, then, we have the eigenket and eigenbra equations for operator $A(t)$,

$$A(t)\big|a_k,t\big\rangle = \big|a_k,t\big\rangle a_k\,, \qquad a_k\big\langle a_k,t\big| = \big\langle a_k,t\big|A(t)\,. \qquad (3.1.3)$$

eigenket
at time t

eigenvalue
that does not
change in time

eigenbra
at time t

operator at time t

They are compactly summarized in the spectral decomposition of $A(t)$

$$A(t) = \sum_{k=1}^{n}\big|a_k,t\big\rangle a_k\big\langle a_k,t\big|\,, \qquad (3.1.4)$$

supplemented by the orthonormality relation

$$\big\langle a_j,t\big|a_k,t\big\rangle = \delta_{jk} \qquad (3.1.5)$$

and the completeness relation

$$\sum_{k=1}^{n}\big|a_k,t\big\rangle\big\langle a_k,t\big| = 1\,. \qquad (3.1.6)$$

These are valid at all times t, but please note that all kets and bras must refer to the same time in these statements.

The time derivative of bra $\big\langle a_k,t\big|$,

$$\frac{\partial}{\partial t}\big\langle a_k,t\big| = \lim_{\tau\to 0}\frac{1}{\tau}\Big(\big\langle a_k,t+\tau\big| - \big\langle a_k,t\big|\Big)\,, \qquad (3.1.7)$$

compares $\big\langle a_k,t\big|$ with $\big\langle a_k,t+\tau\big|$. Since we have an orthonormal, complete set of such bras at each time, they must be related to each other by an infinitesimal unitary transformation when τ is tiny,

$$\big\langle a_k,t+\tau\big| = \big\langle a_k,t\big|U_t(\tau) \qquad (3.1.8)$$

with $[U_t(\tau)]^\dagger U_t(\tau) = 1$ and $U_t(\tau = 0) = 1$. Writing

$$U_t(\tau) = 1 - \frac{\mathrm{i}}{\hbar}H(t)\tau + \underbrace{\cdots}_{\text{terms of order }\tau^2} \qquad (3.1.9)$$

with a constant $\hbar$ of metrical dimension energy $\times$ time, so that operator $H(t)$ has the metrical dimension of energy, we have

$$[U_t(\tau)]^\dagger = 1 + \frac{i}{\hbar} H(t)^\dagger \tau + \cdots \tag{3.1.10}$$

and get

$$[U_t(\tau)]^\dagger U_t(\tau) = 1 + \frac{i}{\hbar} \big[H(t)^\dagger - H(t) \big] \tau + \cdots . \tag{3.1.11}$$

This tells us that $H(t)$ must be hermitian,

$$H(t)^\dagger = H(t) . \tag{3.1.12}$$

It is the hermitian generator of time transformations, the quantum analog of William R. Hamilton's function of classical mechanics. We borrow the terminology and call it the *Hamilton operator*. Such as the classical Hamilton function is the energy of the evolving system, the quantum Hamilton operator is the energy operator. Its eigenvalues are the possible measurement results that one can get when measuring the energy, and the eigenkets and eigenbras of H describe states of the system with definite energy.

The constant $\hbar$, that is needed to render $U_t(\tau)$ dimensionless, is a natural constant. In honor of Max Planck one calls it *Planck's constant*. The numerical value is

$$\hbar = 1.05457 \times 10^{-34}\,\text{Js} . \tag{3.1.13}$$

It is so very small because we use macroscopic units (Joule for energy, seconds for time) to express it. In microscopic units (eV for energy, femtoseconds for time) the value of $\hbar$ is of order unity,

$$\hbar = 0.658212\,\text{eV fs} , \tag{3.1.14}$$

as it should be.

With $U_t(\tau) = 1 - \frac{i}{\hbar} H(t)\tau + \cdots$, we then have

$$\langle a_k, t + \tau | - \langle a_k, t | = \langle a_k, t | (U_t(\tau) - 1)$$
$$= -\frac{i}{\hbar}\tau \langle a_k, t | H(t) + \cdots \tag{3.1.15}$$

and

$$i\hbar \frac{\partial}{\partial t} \langle a_k, t | = \langle a_k, t | H(t) \tag{3.1.16}$$

follows. There is nothing special about operator $A(t)$, just as well the bra
could refer to another operator, so we suppress the quantum number a_k
and write more generally

$$i\hbar \frac{\partial}{\partial t} \langle \ldots, t| = \langle \ldots, t|H(t) , \tag{3.1.17}$$

where the ellipses means the same quantum numbers on both sides. This is
Erwin Schrödinger's equation of motion, the celebrated *Schrödinger equation* — actually the one for bras, which is equivalent to, but not identical
with, the original form of the equation that Schrödinger discovered. The
adjoint of (3.1.17),

$$-i\hbar \frac{\partial}{\partial t} |\ldots, t\rangle = H(t)|\ldots, t\rangle , \tag{3.1.18}$$

is the Schrödinger equation for kets. And there is a plethora of numerical
versions of the Schrödinger equation.

3.2   Heisenberg equation

The time derivative of operator $A(t)$,

$$\begin{aligned}
\frac{\mathrm{d}}{\mathrm{d}t} A(t) &= \sum_k \left(\frac{\partial |a_k, t\rangle}{\partial t} a_k \langle a_k, t| + |a_k, t\rangle a_k \frac{\partial \langle a_k, t|}{\partial t} \right) \\
&= \sum_k \left(-\frac{1}{i\hbar} H(t)|a_k, t\rangle a_k \langle a_k, t| + |a_k, t\rangle a_k \langle a_k, t| \frac{1}{i\hbar} H(t) \right) \\
&= -\frac{1}{i\hbar} H(t)A(t) + \frac{1}{i\hbar} A(t)H(t)
\end{aligned} \tag{3.2.1}$$

is conveniently written as

$$\frac{\mathrm{d}}{\mathrm{d}t} A(t) = \frac{1}{i\hbar} [A(t), H(t)] , \tag{3.2.2}$$

where

$$[X, Y] = XY - YX = -[Y, X] \tag{3.2.3}$$

denotes the so-called *commutator* of operators X and Y. It has a lot in
common with a differentiation. In particular, there is a sum rule

$$\begin{aligned}
[X_1 + X_2, Y] &= [X_1, Y] + [X_2, Y] , \\
[X, Y_1 + Y_2] &= [X, Y_1] + [X, Y_2] ,
\end{aligned} \tag{3.2.4}$$

and a product rule,

$$[X_1 X_2, Y] = [X_1, Y]X_2 + X_1[X_2, Y],$$
$$[X, Y_1 Y_2] = [X, Y_1]Y_2 + Y_1[X, Y_2], \tag{3.2.5}$$

where due attention must be paid to the order in which the operators are appearing in these products. These relations, and many more, are easily verified by inspection.

3-1 Show that

$$[[X, Y], Z] + [[Y, Z], X] + [[Z, X], Y] = 0.$$

This is an example, here for commutators, of a class of equations of a particular structure, which are commonly referred to as *Jacobi identities*, in recognition of Karl G. J. Jacobi's contribution.

A bit more general than $A(t)$ with time-independent eigenvalues, is the case where this restriction is lifted. As a consequence of such a *parametric* time-dependence, we then have an extra term

$$\sum_k |a_k, t\rangle \frac{\partial a_k}{\partial t} \langle a_k, t| \equiv \frac{\partial A}{\partial t} \tag{3.2.6}$$

in

$$\frac{d}{dt} A = \frac{\partial A}{\partial t} + \frac{1}{i\hbar}[A, H], \tag{3.2.7}$$

the *Heisenberg equation of motion*, aptly named in honor of its discoverer Weiner Heisenberg. By contrast, the commutator term $\frac{1}{i\hbar}[A, H]$ originates in physical interactions, as they are contributing various amounts of interaction energy to the Hamilton operator H, so that $\frac{1}{i\hbar}[A, H]$ refers to the *dynamical* time dependence of A.

As a rule, a parametric time dependence results from human interference. For instance, we choose to assign different numbers to the measurement results at different times, because for some reason that is convenient. Or, we actually change a parameter of the apparatus in question, such as altering the direction and strength of the magnetic field that rotates the magnetic moments in the set-ups on pages 81 and 82. The magnetic interaction energy, $-\vec{\mu} \cdot \vec{B}(t)$, then introduces the parametric time dependence into the Hamilton operator. It is, in fact, the only net time dependence that H can possess, because $[H, H] = 0$ holds always, so that Heisenberg's equation says

$$\frac{\mathrm{d}}{\mathrm{d}t}H = \frac{\partial}{\partial t}H \qquad (3.2.8)$$

for the Hamilton operator itself: The time dependence of H is always parametric in nature.

3.3 Equivalent Hamilton operators

In (3.2.7), the Heisenberg equation of motion for A, H enters only in the commutator, and therefore nothing changes if we add a constant ($\equiv$ a multiple of the identity operator) to the Hamilton operator,

$$H \to H + \hbar\Omega : \quad \frac{\mathrm{d}}{\mathrm{d}t}A(t) \quad \text{unchanged.} \qquad (3.3.1)$$

But, there is a change in the Schrödinger equation, so that we must carefully compare

$$i\hbar\frac{\partial}{\partial t}\langle\ldots,t| = \langle\ldots,t|H \qquad \text{(original)} \qquad (3.3.2)$$

with

$$i\hbar\frac{\partial}{\partial t}\langle\ldots,t|' = \langle\ldots,t|'(H + \hbar\Omega) \qquad \text{(altered)} \qquad (3.3.3)$$

where the change of the Hamilton operator is accompanied by a corresponding change of the bras,

$$\langle\ldots,t| \to \langle\ldots,t|'. \qquad (3.3.4)$$

We have

$$\left(i\hbar\frac{\partial}{\partial t} - \hbar\Omega\right)\langle\ldots,t|' = \langle\ldots,t|'H \qquad (3.3.5)$$

and note that the differential operator on the left can be written as

$$i\hbar\frac{\partial}{\partial t} - \hbar\Omega = i\hbar\left(\frac{\partial}{\partial t} + i\Omega\right) = i\hbar\,e^{-i\Omega t}\frac{\partial}{\partial t}\,e^{i\Omega t}. \qquad (3.3.6)$$

(Keep in mind that there is an unwritten function of t to the right of these expressions, the function that is differentiated.) Thus, equivalently we have

$$i\hbar\frac{\partial}{\partial t}\left(e^{i\Omega t}\langle\ldots,t|'\right) = \left(e^{i\Omega t}\langle\ldots,t|'\right)H \qquad (3.3.7)$$

and the identification

$$\langle\ldots,t|' = \mathrm{e}^{-\mathrm{i}\Omega t}\langle\ldots,t| \tag{3.3.8}$$

turn this into the original Schrödinger equation for $\langle\ldots,t|$.

The phase factor $\mathrm{e}^{-\mathrm{i}\Omega t}$ that is acquired by the bra vectors is the *same* for all bras, there is no dependence on the implicit quantum numbers indicated by the ellipsis. As a consequence, all bras are just multiplied by a common phase factor, a freedom we have anyway. Such a phase factor is of no physical relevance, although it might be required for mathematical consistency. Since the kets acquire the complex conjugate phase factor

$$|\ldots,t\rangle \to |\ldots,t\rangle' = |\ldots,t\rangle\,\mathrm{e}^{\mathrm{i}\Omega t}, \tag{3.3.9}$$

no visible phase factors appear in the spectral decomposition of (3.1.4),

$$A = \sum_{k=1}^{n}|a_k,t\rangle a_k\langle a_k,t| = \sum_{k=1}^{n}|a_k,t\rangle' a_k\langle a_k,t|'. \tag{3.3.10}$$

Clearly, there is no physical significance of the replacement $H \to H + \hbar\Omega$. We can redefine the origin of the energy scale anyway we wish: *energy differences* are physically significant, not absolute energy values.

3-2 You can even have a time-dependent shift of the origin on the energy scale,

$$H \to H + \hbar\Omega(t)\,.$$

How do you then relate $\langle\ldots,t|'$ to $\langle\ldots,t|$?

3.4   von Neumann equation

We noted above that the Hamilton operator is particular because it has no dynamical time dependence, only a parametric one — very often none at all, when the external parameters do not change in time. Another special operator is the statistical operator ρ that summarizes our knowledge about the preparation of the system. As such it is the mathematical symbol for phrases like "atoms that are $\uparrow_z$ at time t_0" which, by their very nature, refer to one particular moment in time, namely the instant at which we learned something about the atoms. As a consequence, a statistical operator has no total time dependence, $\dfrac{\mathrm{d}}{\mathrm{d}t}\rho = 0$, it is simply constant in time, as an operator. The Heisenberg equation of motion,

$$0 = \frac{d}{dt}\rho = \frac{\partial}{\partial t}\rho + \frac{1}{i\hbar}[\rho, H], \qquad (3.4.1)$$

then says that the parametric time dependence of ρ — term $\frac{\partial}{\partial t}\rho$ — and the dynamical time dependence — term $\frac{1}{i\hbar}[\rho, H]$ — compensate for each other exactly. In recognition of John von Neumann's contributions, this equation of motion for the statistical operator is often referred to as the *von Neumann equation*; we regard it as a special case of the Heisenberg equation of motion.

3.5 Example: Larmor precession

As an example, let us consider the statistical operator for magnetic silver atoms of (2.15.15),

$$\rho = \frac{1}{2}\left(1 + \vec{s}(t) \cdot \vec{\sigma}(t)\right), \qquad (3.5.1)$$

parametric ⌐ ⌐ dynamical
time dependence time dependence

where we now make explicit that both the Pauli vector operator $\vec{\sigma}$ and its expectation value $\vec{s} = \langle\vec{\sigma}\rangle = \text{tr}\{\rho\vec{\sigma}\}$ depend on time. The time dependence of $\vec{\sigma}(t)$ is dynamical, whereas that of $\vec{s}(t)$ is parametric: a change of $\vec{s}$ changes the functional dependence of ρ on the dynamical variables $\sigma_x, \sigma_y, \sigma_z$.

Here, the two time dependences compensate for each other, so that the product $\vec{s}(t) \cdot \vec{\sigma}(t)$ as a whole is time independent. In particular, then, the eigenvalues of ρ do not change in time, and since they are given by

$$\frac{1}{2} \pm \frac{1}{2}|\vec{s}(t)|, \qquad (3.5.2)$$

the length of the numerical vector $\vec{s}(t)$ is constant in time. Therefore, the only change of $\vec{s}$ from t to $t + dt$ can be an infinitesimal rotation,

$$\frac{d}{dt}\vec{s}(t) = \vec{\omega}(t) \times \vec{s}(t), \qquad (3.5.3)$$

where the angular velocity vector $\vec{\omega}(t)$ could itself depend on time. Since

$\vec{s}(t) = \langle \vec{\sigma}(t) \rangle$, we infer that

$$\frac{d}{dt}\vec{\sigma}(t) = \vec{\omega}(t) \times \vec{\sigma}(t).$$ (3.5.4)

Indeed, then

$$\frac{d}{dt}(\vec{s} \cdot \vec{\sigma}) = \frac{d\vec{s}}{dt} \cdot \vec{\sigma} + \vec{s} \cdot \frac{d\vec{\sigma}}{dt} = (\vec{\omega} \times \vec{s}) \cdot \vec{\sigma} + \vec{s} \cdot (\vec{\omega} \times \vec{\sigma}) = 0.$$ (3.5.5)

The right-hand side of (3.5.4) is the right-hand side of Heisenberg's equation $\frac{d}{dt}\vec{\sigma} = \frac{1}{i\hbar}[\vec{\sigma}, H]$, so that

$$[\vec{\sigma}, H] = i\hbar\vec{\omega} \times \vec{\sigma}.$$ (3.5.6)

The Hamilton operator must be a function of $\vec{\sigma}$, there are no other operators available to construct it, and since all functions of $\vec{\sigma}$ are linear functions, we have

$$H = \hbar\vec{\Omega} \cdot \vec{\sigma}$$ (3.5.7)

with $\vec{\Omega}(t)$ related to $\vec{\omega}(t)$. There is the further possibility of adding a multiple of the identity to H but, as discussed in Section 3.3, this is of no consequence, so we do not have to take it into account.

To find the relation between $\vec{\omega}$ and $\vec{\Omega}$, we recall (2.9.12),

$$\vec{a} \cdot \vec{\sigma}\, \vec{b} \cdot \vec{\sigma} = \vec{a} \cdot \vec{b} + i(\vec{a} \times \vec{b}) \cdot \vec{\sigma}$$ (3.5.8)

and

$$\vec{b} \cdot \vec{\sigma}\, \vec{a} \cdot \vec{\sigma} = \vec{a} \cdot \vec{b} - i(\vec{a} \times \vec{b}) \cdot \vec{\sigma},$$ (3.5.9)

valid for all numerical 3-vectors $\vec{a}$ and $\vec{b}$. Their difference is the commutation relation

$$[\vec{a} \cdot \vec{\sigma}, \vec{b} \cdot \vec{\sigma}] = 2i(\vec{a} \times \vec{b}) \cdot \vec{\sigma} = 2i\vec{a} \cdot (\vec{b} \times \vec{\sigma}),$$ (3.5.10)

so that

$$[\vec{\sigma}, \vec{b} \cdot \vec{\sigma}] = 2i\vec{b} \times \vec{\sigma},$$ (3.5.11)

after getting rid of the arbitrary vector $\vec{a}$. The comparison with

$$[\vec{\sigma}, H] = [\vec{\sigma}, \hbar\vec{\Omega} \cdot \vec{\sigma}] = 2i\hbar\vec{\Omega} \times \vec{\sigma}$$

$$= i\hbar\vec{\omega} \times \vec{\sigma}$$ (3.5.12)

establishes $\vec{\Omega} = \frac{1}{2}\vec{\omega}$, and we arrive at

$$H = \frac{1}{2}\hbar\vec{\omega}\cdot\vec{\sigma}. \tag{3.5.13}$$

3-3 For a silver atom with magnetic moment $\vec{\mu}$, we have the magnetic interaction energy $-\vec{\mu}\cdot\vec{B}$ of (2.1.5). The relation between $\vec{\mu}$ and $\vec{\sigma}$ is

$$\vec{\mu} = -\mu_B\vec{\sigma}$$

where $\mu_B = 9.274 \times 10^{-24}\,\text{J/T}$ (Joule/Tesla) is the so-called *Bohr magneton*, and the negative sign comes about because electrons carry negative charge. How is $\vec{\omega}$ related to $\vec{B}$? How large is $\omega = |\vec{\omega}|$ for a magnetic field of $100\,\text{G} = 0.01\,\text{T}$?

Now, to be more specific, let us consider $\vec{\omega} = \omega\vec{e}_z$ with $\omega > 0$ independent of time t. Then

$$\frac{\mathrm{d}}{\mathrm{d}t}\vec{\sigma} = \omega\vec{e}_z \times \vec{\sigma}, \tag{3.5.14}$$

with components

$$\frac{\mathrm{d}}{\mathrm{d}t}\sigma_x = \vec{e}_x\cdot(\omega\vec{e}_z\times\vec{\sigma}) = \omega\underbrace{(\vec{e}_x\times\vec{e}_z)}_{=-\vec{e}_y}\cdot\vec{\sigma} = -\omega\sigma_y\,,$$

$$\frac{\mathrm{d}}{\mathrm{d}t}\sigma_y = \vec{e}_y\cdot(\omega\vec{e}_z\times\vec{\sigma}) = \omega\underbrace{(\vec{e}_y\times\vec{e}_z)}_{=\vec{e}_x}\cdot\vec{\sigma} = \omega\sigma_x\,,$$

$$\frac{\mathrm{d}}{\mathrm{d}t}\sigma_z = \vec{e}_z\cdot(\omega\vec{e}_z\times\vec{\sigma}) = 0\,, \tag{3.5.15}$$

which are, of course, exactly what we get from Heisenberg's equations directly,

$$\frac{\mathrm{d}}{\mathrm{d}t}\sigma_x = \frac{1}{\mathrm{i}\hbar}[\sigma_x, H] = \frac{1}{2\mathrm{i}}\omega\underbrace{[\sigma_x,\sigma_z]}_{=-2\mathrm{i}\sigma_y} = -\omega\sigma_y\,,$$

$$\frac{\mathrm{d}}{\mathrm{d}t}\sigma_y = \frac{1}{\mathrm{i}\hbar}[\sigma_y, H] = \frac{1}{2\mathrm{i}}\omega\underbrace{[\sigma_y,\sigma_z]}_{=2\mathrm{i}\sigma_x} = \omega\sigma_x\,,$$

$$\frac{\mathrm{d}}{\mathrm{d}t}\sigma_z = \frac{1}{\mathrm{i}\hbar}[\sigma_z, H] = \frac{1}{2\mathrm{i}}\omega\underbrace{[\sigma_z,\sigma_z]}_{=0} = 0\,, \tag{3.5.16}$$

where the Hamilton operator

$$H = \frac{1}{2}\hbar\omega\sigma_z \tag{3.5.17}$$

is obtained from (3.5.13) for $\vec{\omega} = \omega\vec{e}_z$.

We have to solve these equations in order to express $\vec{\sigma}(t)$ in terms of $\vec{\sigma}(0)$. The solution for σ_z is immediate,

$$\sigma_z(t) = \sigma_z(0), \tag{3.5.18}$$

and to find that for σ_x and σ_y, we look at $\sigma_x + i\sigma_y$,

$$\frac{d}{dt}(\sigma_x + i\sigma_y) = -\omega\sigma_y + i\omega\sigma_x = i\omega(\sigma_x + i\sigma_y), \tag{3.5.19}$$

which is solved by

$$\sigma_x(t) + i\sigma_y(t) = e^{i\omega t}\big(\sigma_x(0) + i\sigma_y(0)\big). \tag{3.5.20}$$

Either by considering $\sigma_x - i\sigma_y$, or by just taking the adjoint, we also get

$$\sigma_x(t) - i\sigma_y(t) = e^{-i\omega t}\big(\sigma_x(0) - i\sigma_y(0)\big), \tag{3.5.21}$$

and then

$$\begin{aligned}
\sigma_x(t) &= \sigma_x(0)\cos(\omega t) - \sigma_y(0)\sin(\omega t), \\
\sigma_y(t) &= \sigma_y(0)\cos(\omega t) + \sigma_x(0)\sin(\omega t)
\end{aligned} \tag{3.5.22}$$

by adding and subtracting. It is a matter of inspection to verify that they obey the differential equations, indeed, and reduce to an identity for $t = 0$.

Since $\vec{s} = \langle\vec{\sigma}\rangle$, we also have

$$\begin{aligned}
s_x(t) &= s_x(0)\cos(\omega t) - s_y(0)\sin(\omega t), \\
s_y(t) &= s_y(0)\cos(\omega t) + s_x(0)\sin(\omega t), \\
s_z(t) &= s_z(0),
\end{aligned} \tag{3.5.23}$$

showing once more, now rather explicitly, that 3-vector $\vec{s}$ precesses around the axis specified by $\vec{\omega}$, here the z axis, with angular velocity $\omega = |\vec{\omega}|$. It takes the time $2\pi/\omega$ to complete one revolution, so we have a periodic motion with this period. In the context of Exercise 3-3, this is the so-called *Larmor precession* (Sir Joseph Larmor, that is) of a magnetic moment in a homogeneous magnetic field.

3.6 Time-dependent probability amplitudes

With the Hamilton operator (3.5.17), the Schrödinger equations for eigen-bras of σ_z are

$$i\hbar\frac{\partial}{\partial t}\langle\uparrow_z,t| = \langle\uparrow_z,t|H(t) = \langle\uparrow_z,t|\frac{\hbar}{2}\omega\sigma_z(t) = \frac{\hbar}{2}\omega\langle\uparrow_z,t| \qquad (3.6.1)$$

and, likewise,

$$i\hbar\frac{\partial}{\partial t}\langle\downarrow_z,t| = -\frac{\hbar}{2}\omega\langle\downarrow_z,t|, \qquad (3.6.2)$$

which are solved by

$$\langle\uparrow_z,t| = e^{-\frac{i}{2}\omega t}\langle\uparrow_z,0|,$$
$$\langle\downarrow_z,t| = e^{\frac{i}{2}\omega t}\langle\downarrow_z,0|. \qquad (3.6.3)$$

We use them to find the time-dependent probability amplitudes for a given state ket,

$$| \rangle = |\uparrow_z,0\rangle\alpha_0 + \langle\downarrow_z,0|\beta_0$$
$$= |\uparrow_z,t\rangle\alpha(t) + |\downarrow_z,t\rangle\beta(t), \qquad (3.6.4)$$

where

$$\alpha(t) = \langle\uparrow_z,t| \rangle = e^{-\frac{i}{2}\omega t}\langle\uparrow_z,0| \rangle = e^{-\frac{i}{2}\omega t}\alpha_0 \qquad (3.6.5)$$

and

$$\beta(t) = \langle\downarrow_z,t| \rangle = e^{\frac{i}{2}\omega t}\langle\downarrow_z,0| \rangle = e^{\frac{i}{2}\omega t}\beta_0. \qquad (3.6.6)$$

As a check of consistency, let us verify that they give us the same expectation values of $\sigma_x, \sigma_y, \sigma_z$ as a function of t. We recall the relations of Exercise 2-27,

$$\langle\sigma_x\rangle = 2\,\mathrm{Re}(\alpha^*\beta), \quad \langle\sigma_y\rangle = 2\,\mathrm{Im}(\alpha^*\beta), \quad \langle\sigma_z\rangle = |\alpha|^2 - |\beta|^2, \quad (3.6.7)$$

and thus find

$$\langle\sigma_x(t)\rangle = 2\,\mathrm{Re}\left(e^{i\omega t}\alpha_0^*\beta_0\right). \qquad (3.6.8)$$

With Euler's identity $e^{i\omega t} = \cos(\omega t) + i\sin(\omega t)$ and the initial values

$$\alpha_0^*\beta_0 = \frac{1}{2}\langle\sigma_x(0)\rangle + \frac{i}{2}\langle\sigma_y(0)\rangle, \qquad (3.6.9)$$

this becomes

$$\langle \sigma_x(t) \rangle = \text{Re}\Big(\big[\cos(\omega t) + \text{i}\sin(\omega t)\big]\big[\langle \sigma_x(0) \rangle + \text{i}\langle \sigma_y(0) \rangle\big]\Big)$$
$$= \langle \sigma_x(0) \rangle \cos(\omega t) - \langle \sigma_y(0) \rangle \sin(\omega t), \qquad (3.6.10)$$

and likewise we get

$$\langle \sigma_y(t) \rangle = \text{Im}\Big(\big[\cos(\omega t) + \text{i}\sin(\omega t)\big]\big[\langle \sigma_x(0) \rangle + \text{i}\langle \sigma_y(0) \rangle\big]\Big)$$
$$= \langle \sigma_y(0) \rangle \cos(\omega t) + \langle \sigma_x(0) \rangle \sin(\omega t), \qquad (3.6.11)$$

and finally,

$$\langle \sigma_z(t) \rangle = \left| e^{-\frac{\text{i}}{2}\omega t}\alpha_0 \right|^2 - \left| e^{\frac{\text{i}}{2}\omega t}\beta_0 \right|^2$$
$$= |\alpha_0|^2 - |\beta_0|^2 = \langle \sigma_z(0) \rangle. \qquad (3.6.12)$$

These are, of course, just the equations for the components of $\vec{s}(t)$ in (3.5.23).

3-4 For $H = \frac{1}{2}\hbar\omega\sigma_z$, state and solve the Schrödinger equations for $\langle \uparrow_x, t|$ and $\langle \downarrow_x, t|$.

3.7   Schrödinger equation for probability amplitudes

Lifting the restriction of $\vec{\omega} = \omega\vec{e}_z$, we return to the more general case (3.5.13) of an arbitrary, perhaps time-dependent, angular velocity vector $\vec{\omega}$,

$$H = \frac{1}{2}\hbar\vec{\omega} \cdot \vec{\sigma}. \qquad (3.7.1)$$

The standard matrix representation of H, which refers to eigenkets and eigenbras of $\sigma_z(t)$, reads

$$H = \frac{\hbar}{2}\vec{\omega} \cdot \vec{\sigma}(t) = \Big(|\uparrow_z, t\rangle, |\downarrow_z, t\rangle \Big) \mathcal{H} \begin{pmatrix} \langle \uparrow_z, t| \\ \langle \downarrow_z, t| \end{pmatrix} \qquad (3.7.2)$$

with

$$\mathcal{H} = \begin{pmatrix} \langle \uparrow_z, t| \\ \langle \downarrow_z, t| \end{pmatrix} H \left(|\uparrow_z, t\rangle, |\downarrow_z, t\rangle \right)$$

$$= \begin{pmatrix} \langle \uparrow_z, t|H|\uparrow_z, t\rangle & \langle \uparrow_z, t|H|\downarrow_z, t\rangle \\ \langle \downarrow_z, t|H|\uparrow_z, t\rangle & \langle \downarrow_z, t|H|\downarrow_z, t\rangle \end{pmatrix}$$

$$= \frac{\hbar}{2} \begin{pmatrix} \omega_z & \omega_x - i\omega_y \\ \omega_x + i\omega_y & -\omega_z \end{pmatrix}. \tag{3.7.3}$$

This matrix appears in the Schrödinger equation obeyed by the column of probability amplitudes, as specified by

$$| \rangle = |\uparrow_z, t\rangle \alpha(t) + |\downarrow_z, t\rangle \beta(t) \mathrel{\widehat{=}} \begin{pmatrix} \alpha(t) \\ \beta(t) \end{pmatrix}, \tag{3.7.4}$$

inasmuch as

$$i\hbar \frac{\partial}{\partial t} \begin{pmatrix} \alpha(t) \\ \beta(t) \end{pmatrix} = i\hbar \frac{\partial}{\partial t} \begin{pmatrix} \langle \uparrow_z, t| \\ \langle \downarrow_z, t| \end{pmatrix} | \rangle$$

$$= \begin{pmatrix} \langle \uparrow_z, t| \\ \langle \downarrow_z, t| \end{pmatrix} H | \rangle$$

$$= \mathcal{H} \begin{pmatrix} \langle \uparrow_z, t| \\ \langle \downarrow_z, t| \end{pmatrix} | \rangle = \mathcal{H} \begin{pmatrix} \alpha(t) \\ \beta(t) \end{pmatrix}, \tag{3.7.5}$$

that is

$$i\hbar \frac{\partial}{\partial t} \begin{pmatrix} \alpha \\ \beta \end{pmatrix} = \mathcal{H} \begin{pmatrix} \alpha \\ \beta \end{pmatrix}. \tag{3.7.6}$$

Please note that in relations such as

$$\begin{pmatrix} \langle \uparrow_z, t| \\ \langle \downarrow_z, t| \end{pmatrix} H = \mathcal{H} \begin{pmatrix} \langle \uparrow_z, t| \\ \langle \downarrow_z, t| \end{pmatrix} \tag{3.7.7}$$

the operator H acts from the right on the bras, whereas its numerical 2×2 matrix $\mathcal{H}$ multiplies the column from the left. In other words, both the operator and the matrix act from their natural sides on the column of the bras.

As an example, consider

$$H = \frac{\hbar}{2}\omega \sigma_x \mathrel{\widehat{=}} \mathcal{H} = \frac{\hbar}{2}\omega \begin{pmatrix} 0 & 1 \\ 1 & 0 \end{pmatrix}, \tag{3.7.8}$$

for which

$$i\hbar\frac{\partial}{\partial t}\begin{pmatrix}\alpha\\\beta\end{pmatrix} = \frac{\hbar}{2}\omega\begin{pmatrix}0 & 1\\1 & 0\end{pmatrix}\begin{pmatrix}\alpha\\\beta\end{pmatrix} = \frac{\hbar}{2}\omega\begin{pmatrix}\beta\\\alpha\end{pmatrix} \qquad (3.7.9)$$

or

$$\frac{\partial}{\partial t}\alpha = -\frac{i}{2}\omega\beta, \qquad \frac{\partial}{\partial t}\beta = -\frac{i}{2}\omega\alpha. \qquad (3.7.10)$$

These simplify if we take the sum and difference of α and β,

$$\frac{\partial}{\partial t}(\alpha+\beta) = -\frac{i\omega}{2}(\alpha+\beta),$$
$$\frac{\partial}{\partial t}(\alpha-\beta) = \frac{i\omega}{2}(\alpha-\beta). \qquad (3.7.11)$$

Their solutions

$$\alpha(t)+\beta(t) = e^{-i\omega t/2}(\alpha_0+\beta_0),$$
$$\alpha(t)-\beta(t) = e^{i\omega t/2}(\alpha_0-\beta_0), \qquad (3.7.12)$$

tell us that

$$\alpha(t) = \alpha_0\cos\frac{\omega t}{2} - i\beta_0\sin\frac{\omega t}{2},$$
$$\beta(t) = \beta_0\cos\frac{\omega t}{2} - i\alpha_0\sin\frac{\omega t}{2}. \qquad (3.7.13)$$

One verifies by inspection that they obey the coupled differential equations of (3.7.10), and reduce to an identity for $t = 0$, when $\alpha(0) = \alpha_0$ and $\beta(0) = \beta_0$.

This pattern is more generally true. If we have n different measurement results for generic operators,

$$A(t)|a_k,t\rangle = |a_k,t\rangle a_k \qquad \text{for} \quad k = 1,2,\ldots,n, \qquad (3.7.14)$$

and consider the Schrödinger equation for the column of bras $\langle a_k,t|$,

$$i\hbar\frac{\partial}{\partial t}\begin{pmatrix}\langle a_1,t|\\\vdots\\\langle a_n,t|\end{pmatrix} = \begin{pmatrix}\langle a_1,t|\\\vdots\\\langle a_n,t|\end{pmatrix}H = \mathcal{H}\begin{pmatrix}\langle a_1,t|\\\vdots\\\langle a_n,t|\end{pmatrix}, \qquad (3.7.15)$$

we encounter the $n \times n$ matrix

$$\mathcal{H} = \begin{pmatrix} \langle a_1, t | H | a_1, t \rangle & \cdots & \langle a_1, t | H | a_n, t \rangle \\ \langle a_2, t | H | a_1, t \rangle & \cdots & \langle a_2, t | H | a_n, t \rangle \\ \vdots & \ddots & \vdots \\ \langle a_n, t | H | a_1, t \rangle & \cdots & \langle a_n, t | H | a_n, t \rangle \end{pmatrix}, \tag{3.7.16}$$

so that the probability amplitudes of an arbitrary ket

$$| \ \rangle = |a_1, t\rangle \alpha_1(t) + |a_2, t\rangle \alpha_2(t) + \cdots + |a_n, t\rangle \alpha_n(t) \cong \begin{pmatrix} \alpha_1(t) \\ \alpha_2(t) \\ \vdots \\ \alpha_n(t) \end{pmatrix} \tag{3.7.17}$$

obey the Schrödinger equation

$$i\hbar \frac{\partial}{\partial t} \begin{pmatrix} \alpha_1(t) \\ \alpha_2(t) \\ \vdots \\ \alpha_n(t) \end{pmatrix} = \mathcal{H} \begin{pmatrix} \alpha_1(t) \\ \alpha_2(t) \\ \vdots \\ \alpha_n(t) \end{pmatrix}. \tag{3.7.18}$$

3-5 For $\mathcal{H} = \hbar\omega \begin{pmatrix} 0 & 0 & 1 \\ 0 & 2 & 0 \\ 1 & 0 & 0 \end{pmatrix}$ and $\begin{pmatrix} \alpha_1(0) \\ \alpha_2(0) \\ \alpha_3(0) \end{pmatrix} = \frac{1}{3} \begin{pmatrix} 1 \\ 2 \\ 2 \end{pmatrix}$, find $\begin{pmatrix} \alpha_1(t) \\ \alpha_2(t) \\ \alpha_3(t) \end{pmatrix}$.

3-6 For $\mathcal{H} = \hbar\omega \begin{pmatrix} 1 & 1 & 0 \\ 1 & 1 & 0 \\ 0 & 0 & -1 \end{pmatrix}$ and $\begin{pmatrix} \alpha_1(0) \\ \alpha_2(0) \\ \alpha_3(0) \end{pmatrix} = \frac{1}{\sqrt{3}} \begin{pmatrix} 1 \\ 1 \\ 1 \end{pmatrix}$, find $\begin{pmatrix} \alpha_1(t) \\ \alpha_2(t) \\ \alpha_3(t) \end{pmatrix}$.

One often writes $\psi(t)$ for such a column of probability amplitudes,

$$\psi(t) = \begin{pmatrix} \alpha_1(t) \\ \vdots \\ \alpha_n(t) \end{pmatrix}, \tag{3.7.19}$$

and then this Schrödinger equation acquires a form,

$$i\hbar \frac{\partial}{\partial t} \psi = \mathcal{H}\psi, \tag{3.7.20}$$

that is as compact as the original statement (3.1.17) about bras.

3.8 Time-independent Schrödinger equation

If $\mathcal{H}$ has no parametric time dependence and is, therefore, constant in time, it is easy to integrate these Schrödinger equations formally, with the outcome

$$\langle\ldots,t| = \langle\ldots,0| \, e^{-iHt/\hbar} \tag{3.8.1}$$

for the original bra equation (3.1.17), and

$$\psi(t) = e^{-i\mathcal{H}t/\hbar}\psi(0) \tag{3.8.2}$$

for its numerical analog (3.7.20) for columns of probability amplitudes. Again, do not fail to note that the operator rightly stands to the right of the bra, whereas the matrix stands to the left of the column.

Since H is hermitian, $H = H^\dagger$, the operator $e^{-iHt/\hbar}$ is unitary, as it must be to conserve the unit length of the bra. But unless one can do something more explicit about this exponential function of the Hamilton operator, this formal solution of the Schrödinger equation is not of much help. Now, if we know the eigenkets and eigenbras of H,

$$H(t)\big|E_k,t\big\rangle = \big|E_k,t\big\rangle E_k\,, \qquad \big\langle E_k,t\big|H(t) = E_k\big\langle E_k,t\big|\,, \tag{3.8.3}$$

labeled by the eigenvalues E_k, the *eigenenergies* of the system, we can write

$$f(H) = \sum_k \big|E_k,t\big\rangle f(E_k)\big\langle E_k,t\big| \tag{3.8.4}$$

for any function of H, in particular that unitary exponential function. In (3.8.3), $H(t)$ is actually independent of time t, and its eigenvalues E_k are constant in time as well.

Regarding the time dependence of the eigenbras and eigenkets of H, we note that it is elementary, because

$$i\hbar\frac{\partial}{\partial t}\big\langle E_k,t\big| = \big\langle E_k,t\big|H = E_k\big\langle E_k,t\big| \tag{3.8.5}$$

gives

$$\big\langle E_k,t\big| = e^{-iE_kt/\hbar}\big\langle E_k,0\big| \tag{3.8.6}$$

and

$$\big|E_k,t\big\rangle = \big|E_k,0\big\rangle \, e^{iE_kt/\hbar} \tag{3.8.7}$$

immediately. As a consequence, the projector $|E_k, t\rangle\langle E_k, t|$ does not depend on time at all, so that we can just use it at some reference time, which we leave implicit, any instant can serve the purpose equally well. So, just writing $|E_k\rangle\langle E_k|$ for $|E_k, t\rangle\langle E_k, t|$, we have

$$f(H) = \sum_k |E_k\rangle f(E_k)\langle E_k| \qquad (3.8.8)$$

with

$$H|E_k\rangle = |E_k\rangle E_k \qquad (3.8.9)$$

and arrive at

$$\mathrm{e}^{-\mathrm{i}Ht/\hbar} = \sum_k |E_k\rangle \, \mathrm{e}^{-\mathrm{i}E_k t/\hbar} \langle E_k| \, . \qquad (3.8.10)$$

Quite analogously, we have

$$\mathrm{e}^{-\mathrm{i}\mathcal{H}t/\hbar} = \sum_k \phi_k \, \mathrm{e}^{-\mathrm{i}E_k t/\hbar} \phi_k^\dagger \qquad (3.8.11)$$

where ϕ_k is the column representing ket $|E_k\rangle$ and its adjoint $\phi_k^\dagger$ is the row for bra $\langle E_k|$. They are, of course, the respective eigencolumns and eigenrows of the matrix $\mathcal{H}$,

$$\mathcal{H}\phi_k = \phi_k E_k \, , \qquad \phi_k^\dagger \mathcal{H} = E_k \phi_k^\dagger \, , \qquad (3.8.12)$$

which are the numerical analogs of (3.8.3).

3-7 For the 3×3 matrix $\mathcal{H}$ in Exercise 3-5 on page 97, find the eigenvalues, eigencolumns, and eigenrows, and use them to construct the unitary matrix $\mathrm{e}^{-\mathrm{i}\mathcal{H}t/\hbar}$ explicitly.

3-8 Apply your result to the initial column $\psi_0 = \dfrac{1}{3}\begin{pmatrix} 1 \\ 2 \\ 2 \end{pmatrix}$ and compare the outcome for $\psi(t)$ with what you found in Exercise 3-5 by solving the differential equation.

3-9 Do this also for the matrix $\mathcal{H}$ and the initial column in Exercise 3-6.

The eigenvalue equation

$$H|E_k\rangle = |E_k\rangle E_k \qquad (3.8.13)$$

is called *time-independent Schrödinger equation*. It is a central problem to determine the eigenvalues and eigenstates of the dynamics of a system, because the energy eigenvalues determine the frequencies that govern the evolution. It is clear on the outset that the E_ks themselves are of no particular physical significance, because we can always change their value by adding a constant to the Hamilton operator; see Section 3.3. Only energy differences have physical meaning.

To illustrate this point once more, consider a system prepared such that its properties are described by the ket

$$| \; \rangle = \sum_k |E_k\rangle \alpha_k \qquad (3.8.14)$$

and we are asking for the probability of finding eigenvalue a_j of observable A at time t. This probability is the absolute square of the probability amplitude

$$
\begin{aligned}
\langle a_j, t| \; \rangle &= \langle a_j, 0| \, e^{-iHt/\hbar} | \; \rangle \\
&= \sum_k \langle a_j, 0| \, e^{-iHt/\hbar} |E_k\rangle \alpha_k \\
&= \sum_k \langle a_j, 0|E_k\rangle \, e^{-iE_k t/\hbar} \alpha_k \, , \qquad (3.8.15)
\end{aligned}
$$

that is:

$$
\begin{aligned}
\left|\langle a_j, t| \; \rangle\right|^2 &= \sum_{k,l} \alpha_l^* \, e^{iE_l t/\hbar} \langle E_l|a_j,0\rangle\langle a_j,0|E_k\rangle \, e^{-iE_k t/\hbar} \alpha_k \\
&= \sum_{k,l} \alpha_l^* \langle E_l|a_j,0\rangle\langle a_j,0|E_k\rangle \alpha_k \, e^{-i(E_k - E_l)t/\hbar} . \qquad (3.8.16)
\end{aligned}
$$

We read off that the (circular) frequencies of the system are

$$\omega_{kl} = \frac{E_k - E_l}{\hbar} \, , \qquad (3.8.17)$$

which appear in the periodic functions

$$e^{-i(E_k - E_l)t/\hbar} = e^{-i\omega_{kl}t} = \cos(\omega_{kl}t) - i\sin(\omega_{kl}t) . \qquad (3.8.18)$$

Since they refer to the difference in energy between two states of definite energy, the ω_{kl} are often called *transition frequencies*.

3-10 Which transition frequency corresponds to an energy difference of $1\,\mathrm{eV}$?

3.9 Example: Two magnetic silver atoms

As an illustrating example, of quite some physical relevance for certain modern applications, consider two magnetic silver atoms, located on the z axis, a fixed distance r apart:

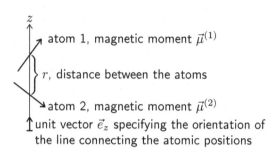

Their magnetic interaction energy is the dipole-dipole interaction energy

$$E_{\text{magn}} = \frac{\vec{\mu}^{(1)} \cdot \vec{\mu}^{(2)} - 3\vec{\mu}^{(1)} \cdot \vec{e}_z \, \vec{e}_z \cdot \vec{\mu}^{(2)}}{r^3} \, . \tag{3.9.1}$$

With $\vec{\mu}^{(1)} \propto \vec{\sigma}^{(1)}$, $\vec{\mu}^{(2)} \propto \vec{\sigma}^{(2)}$, this gives the Hamilton operator

$$H = \frac{1}{2}\hbar\omega \left(\vec{\sigma}^{(1)} \cdot \vec{\sigma}^{(2)} - 3\sigma_z^{(1)}\sigma_z^{(2)} \right) , \tag{3.9.2}$$

where the prefactor $\frac{1}{2}\hbar\omega$, of metrical dimension energy, summarizes the various constants in a convenient way. The energy unit $\hbar\omega$ introduces a corresponding frequency unit ω — or rather $\omega/(2\pi)$, because ω itself is a circular frequency, whereas $\omega/(2\pi)$ is a true frequency.

The dipole-dipole energy E_{magn} is such that the "head to tail" configurations,

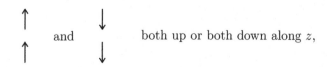

both up or both down along z,

are of minimal energy, $1 - 3 = -2$ units of energy, whereas the "head to

head" or "tail to tail" configurations,

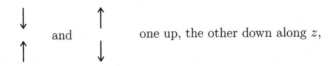

are of much larger energy: $-1 + 3 = +2$ units. Accordingly, we expect $\uparrow\uparrow$ and $\downarrow\downarrow$ to be stable configurations, and $\uparrow\downarrow$ or $\downarrow\uparrow$ to be unstable.

These considerations invite the use of $|\uparrow_z\uparrow_z\rangle, |\uparrow_z\downarrow_z\rangle, |\downarrow_z\uparrow_z\rangle, |\downarrow_z\downarrow_z\rangle$ as the reference kets to which we refer all kets, operators, etc. Thus,

$$| \rangle = \left(|\uparrow_z\uparrow_z\rangle, |\uparrow_z\downarrow_z\rangle, |\downarrow_z\uparrow_z\rangle, |\downarrow_z\downarrow_z\rangle \right) \underbrace{\begin{pmatrix} \psi_1 \\ \psi_2 \\ \psi_3 \\ \psi_4 \end{pmatrix}}_{\widehat{=}\,\psi} \widehat{=} \psi \qquad (3.9.3)$$

for any arbitrary ket, with the particular examples,

$$|\uparrow_z\uparrow_z\rangle \widehat{=} \begin{pmatrix} 1 \\ 0 \\ 0 \\ 0 \end{pmatrix}, \qquad |\uparrow_z\downarrow_z\rangle \widehat{=} \begin{pmatrix} 0 \\ 1 \\ 0 \\ 0 \end{pmatrix},$$

$$|\downarrow_z\uparrow_z\rangle \widehat{=} \begin{pmatrix} 0 \\ 0 \\ 1 \\ 0 \end{pmatrix}, \qquad |\downarrow_z\downarrow_z\rangle \widehat{=} \begin{pmatrix} 0 \\ 0 \\ 0 \\ 1 \end{pmatrix}, \qquad (3.9.4)$$

for the chosen set of basis kets. All of them are eigenkets of $\sigma_z^{(1)}\sigma_z^{(2)}$ with eigenvalues of $(+1)(+1) = +1$, $(+1)(-1) = -1$, $(-1)(+1) = -1$, $(-1)(-1) = +1$, respectively, so that,

$$\sigma_z^{(1)}\sigma_z^{(2)} \widehat{=} \begin{pmatrix} 1 & 0 & 0 & 0 \\ 0 & -1 & 0 & 0 \\ 0 & 0 & -1 & 0 \\ 0 & 0 & 0 & 1 \end{pmatrix} \qquad (3.9.5)$$

is the resulting matrix representation for $\sigma_z^{(1)}\sigma_z^{(2)}$. Regarding $\vec{\sigma}^{(1)} \cdot \vec{\sigma}^{(2)}$, we recall what we found in Section 2.18, namely that

$$\frac{1}{2}\Big(|\uparrow_z\downarrow_z\rangle - |\downarrow_z\uparrow_z\rangle\Big)\Big(\langle\uparrow_z\downarrow_z| - \langle\downarrow_z\uparrow_z|\Big) = \frac{1}{4}\big(1 - \vec{\sigma}^{(1)}\cdot\vec{\sigma}^{(2)}\big), \qquad (3.9.6)$$

so that

$$\vec{\sigma}^{(1)}\cdot\vec{\sigma}^{(2)} = 1 - 2\Big(|\uparrow_z\downarrow_z\rangle - |\downarrow_z\uparrow_z\rangle\Big)\Big(\langle\uparrow_z\downarrow_z| - \langle\downarrow_z\uparrow_z|\Big)$$

$$\widehat{=} \begin{pmatrix} 1 & 0 & 0 & 0 \\ 0 & 1 & 0 & 0 \\ 0 & 0 & 1 & 0 \\ 0 & 0 & 0 & 1 \end{pmatrix} - 2\begin{pmatrix} 0 \\ 1 \\ -1 \\ 0 \end{pmatrix}\begin{pmatrix} 0 & 1 & -1 & 0 \end{pmatrix}$$

$$= \begin{pmatrix} 1 & 0 & 0 & 0 \\ 0 & -1 & 2 & 0 \\ 0 & 2 & -1 & 0 \\ 0 & 0 & 0 & 1 \end{pmatrix}, \qquad (3.9.7)$$

is the matrix representation for $\vec{\sigma}^{(1)}\cdot\vec{\sigma}^{(2)}$.

Putting the ingredients together, we have

$$H \widehat{=} \mathcal{H} = \frac{1}{2}\hbar\omega\left[\begin{pmatrix} 1 & 0 & 0 & 0 \\ 0 & -1 & 2 & 0 \\ 0 & 2 & -1 & 0 \\ 0 & 0 & 0 & 1 \end{pmatrix} - 3\begin{pmatrix} 1 & 0 & 0 & 0 \\ 0 & -1 & 0 & 0 \\ 0 & 0 & -1 & 0 \\ 0 & 0 & 0 & 1 \end{pmatrix}\right]$$

$$= \hbar\omega\begin{pmatrix} -1 & 0 & 0 & 0 \\ 0 & 1 & 1 & 0 \\ 0 & 1 & 1 & 0 \\ 0 & 0 & 0 & -1 \end{pmatrix} \qquad (3.9.8)$$

for the matrix representing the Hamilton operator (3.9.2). The time-independent Schrödinger equation $H|E_k\rangle = |E_k\rangle E_k$ translates into the matrix eigenvalue equation

$$\mathcal{H}\psi = E\psi, \qquad (3.9.9)$$

the solution of which can be found by simple inspection:

$$
\mathcal{H}\begin{pmatrix}1\\0\\0\\0\end{pmatrix} = -\hbar\omega\begin{pmatrix}1\\0\\0\\0\end{pmatrix} : \text{eigenvalue } -\hbar\omega, \quad \text{eigencolumn } \begin{pmatrix}1\\0\\0\\0\end{pmatrix},
$$

$$
\mathcal{H}\begin{pmatrix}0\\0\\0\\1\end{pmatrix} = -\hbar\omega\begin{pmatrix}0\\0\\0\\1\end{pmatrix} : \text{eigenvalue } -\hbar\omega, \quad \text{eigencolumn } \begin{pmatrix}0\\0\\0\\1\end{pmatrix},
$$

$$
\mathcal{H}\begin{pmatrix}0\\1\\1\\0\end{pmatrix} = 2\hbar\omega\begin{pmatrix}0\\1\\1\\0\end{pmatrix} : \text{eigenvalue } 2\hbar\omega, \text{ eigencolumn } \frac{1}{\sqrt{2}}\begin{pmatrix}0\\1\\1\\0\end{pmatrix},
$$

$$
\mathcal{H}\begin{pmatrix}0\\1\\-1\\0\end{pmatrix} = 0 : \qquad \text{eigenvalue } 0, \quad \text{eigencolumn } \frac{1}{\sqrt{2}}\begin{pmatrix}0\\1\\-1\\0\end{pmatrix}.
$$

$$(3.9.10)$$

The factors of $1/\sqrt{2}$ are for the normalization to unit length.

The general statement (3.8.8) about a function of operator H has the numerical analog

$$
f(\mathcal{H}) = \sum_k \psi_{E_k} f(E_k) \psi_{E_k}^{\dagger} \tag{3.9.11}
$$

where ψ_{E_k} is the column for ket $\left|E_k\right\rangle$ and its adjoint $\psi_{E_k}^{\dagger}$ is the row for bra $\left\langle E_k\right|$. In the present example, we thus have

$$
f(\mathcal{H}) = f(-\hbar\omega)\left[\begin{pmatrix}1\\0\\0\\0\end{pmatrix}\begin{pmatrix}1&0&0&0\end{pmatrix} + \begin{pmatrix}0\\0\\0\\1\end{pmatrix}\begin{pmatrix}0&0&0&1\end{pmatrix}\right]
$$

$$
+ f(2\hbar\omega)\frac{1}{2}\begin{pmatrix}0\\1\\1\\0\end{pmatrix}\begin{pmatrix}0&1&1&0\end{pmatrix} + f(0)\frac{1}{2}\begin{pmatrix}0\\1\\-1\\0\end{pmatrix}\begin{pmatrix}0&1&-1&0\end{pmatrix}
$$

$$(3.9.12)$$

or

$$f(\mathcal{H}) = \begin{pmatrix} f(-\hbar\omega) & 0 & 0 & 0 \\ 0 & \frac{1}{2}[f(0) + f(2\hbar\omega)] & \frac{1}{2}[-f(0) + f(2\hbar\omega)] & 0 \\ 0 & \frac{1}{2}[-f(0) + f(2\hbar\omega)] & \frac{1}{2}[f(0) + f(2\hbar\omega)] & 0 \\ 0 & 0 & 0 & f(-\hbar\omega) \end{pmatrix}.$$

$$(3.9.13)$$

We need it for $f(\mathcal{H}) = e^{-i\mathcal{H}t/\hbar}$, that is

$$f(-\hbar\omega) = e^{i\omega t},$$

$$\frac{1}{2}[f(0) + f(2\hbar\omega)] = e^{-i\omega t}\cos(\omega t),$$

$$\frac{1}{2}[-f(0) + f(2\hbar\omega)] = -i\,e^{-i\omega t}\sin(\omega t),\qquad (3.9.14)$$

thus

$$e^{-iHt/\hbar} \cong e^{-i\mathcal{H}t/\hbar} = \begin{pmatrix} e^{i\omega t} & 0 & 0 & 0 \\ 0 & e^{-i\omega t}\cos(\omega t) & -i\,e^{-i\omega t}\sin(\omega t) & 0 \\ 0 & -i\,e^{-i\omega t}\sin(\omega t) & e^{-i\omega t}\cos(\omega t) & 0 \\ 0 & 0 & 0 & e^{i\omega t} \end{pmatrix}.$$

$$(3.9.15)$$

In particular, then, we get

$$\psi(t=0) = \begin{pmatrix} 1 \\ 0 \\ 0 \\ 0 \end{pmatrix} \to \psi(t) = e^{-i\mathcal{H}t/\hbar}\begin{pmatrix} 1 \\ 0 \\ 0 \\ 0 \end{pmatrix} = \begin{pmatrix} e^{i\omega t} \\ 0 \\ 0 \\ 0 \end{pmatrix} = e^{i\omega t}\psi(0),$$

$$(3.9.16)$$

which, translated back into a statement about kets, reads

$$\underbrace{\left|\uparrow_z\uparrow_z, t=0\right\rangle}_{\substack{\text{time-indepen-}\\\text{dent state ket}}} = \underbrace{\left|\uparrow_z\uparrow_z, t\right\rangle}_{\substack{\text{time-de-}\\\text{pendent}\\\text{basis ket}}} \underbrace{e^{i\omega t}}_{\substack{\text{time-dependent}\\\text{probability}\\\text{amplitude}}} \qquad (3.9.17)$$

where we emphasize once more that the ket on the left, which specifies the state of the two-atom system, refers to the initial time $t = 0$ and, therefore, has no actual time dependence, and that the time dependence of the ket and probability amplitude on the right compensate for each other.

Thus atoms that are initially aligned, $\uparrow_z\uparrow_z$, remain aligned as time passes. The same is true for $\downarrow_z\downarrow_z$, the other stable configuration, inasmuch as

$$\psi(0) = \begin{pmatrix} 0 \\ 0 \\ 0 \\ 1 \end{pmatrix} \rightarrow \psi(t) = e^{-i\mathcal{H}t/\hbar}\psi(0) = \begin{pmatrix} 0 \\ 0 \\ 0 \\ e^{i\omega t} \end{pmatrix} = e^{i\omega t}\psi(0)\,. \quad (3.9.18)$$

But the unstable configurations, $\uparrow_z\downarrow_z$ and $\downarrow_z\uparrow_z$, do evolve in time into other arrangements. For example, if we have $\uparrow_z\downarrow_z$ initially, that is

$$\psi(0) = \begin{pmatrix} 0 \\ 1 \\ 0 \\ 0 \end{pmatrix}, \quad (3.9.19)$$

we get

$$\psi(t) = e^{-i\mathcal{H}t/\hbar}\psi(0) = e^{-i\omega t}\begin{pmatrix} 0 \\ \cos(\omega t) \\ -i\sin(\omega t) \\ 0 \end{pmatrix}, \quad (3.9.20)$$

stating that the probability of $\uparrow_z\downarrow_z$ at time t is

$$\left| e^{-i\omega t}\cos(\omega t) \right|^2 = \cos(\omega t)^2\,, \quad (3.9.21)$$

and that for $\downarrow_z\uparrow_z$ at time t is

$$\left| e^{-i\omega t}(-i)\sin(\omega t) \right|^2 = \sin(\omega t)^2\,, \quad (3.9.22)$$

whereas the probabilities of $\uparrow_z\uparrow_z$ and $\downarrow_z\downarrow_z$ are zero for all times. So, if we wait for time $t = \pi/(2\omega)$, then $\uparrow_z\downarrow_z$ has been completely turned into $\downarrow_z\uparrow_z$, and waiting for another period $\pi/(2\omega)$, thus reaching $t = \pi/\omega$, turns it back into $\uparrow_z\downarrow_z$. There is a periodic transition from $\uparrow_z\downarrow_z$ to $\downarrow_z\uparrow_z$ and back, with superpositions at intermediate times. For instance, at time $t = \pi/(4\omega)$, we have

$$\psi\left(t = \frac{\pi}{4\omega}\right) = e^{-i\frac{\pi}{4}}\begin{pmatrix} 0 \\ \cos\frac{\pi}{4} \\ -i\sin\frac{\pi}{4} \\ 0 \end{pmatrix} = \frac{1-i}{2}\begin{pmatrix} 0 \\ 1 \\ -i \\ 0 \end{pmatrix}, \quad (3.9.23)$$

which translates into

$$\left|\uparrow_z\downarrow_z, 0\right\rangle = \left(\left|\uparrow_z\downarrow_z, \frac{\pi}{4\omega}\right\rangle - i\left|\downarrow_z\uparrow_z, \frac{\pi}{4\omega}\right\rangle\right)\frac{1-i}{2}\,. \quad (3.9.24)$$

The frequency at which these oscillations happen is revealed upon writing the probabilities as

$$\left.\begin{array}{c}\cos(\omega t)^2 \\ \sin(\omega t)^2\end{array}\right\} = \frac{1}{2} \pm \frac{1}{2}\cos(2\omega t) \tag{3.9.25}$$

so that $2\omega = (2\hbar\omega - 0\hbar\omega)/\hbar$ is the transition frequency, correctly related to the energy difference between the eigenstates of H that are involved, see

$$|\uparrow_z\downarrow_z\rangle = \frac{1}{\sqrt{2}}\left[\underbrace{\frac{1}{\sqrt{2}}\Big(|\uparrow_z\downarrow_z\rangle + |\downarrow_z\uparrow_z\rangle\Big)}_{\text{energy eigenvalue } 2\hbar\omega} + \underbrace{\frac{1}{\sqrt{2}}\Big(|\uparrow_z\downarrow_z\rangle - |\downarrow_z\uparrow_z\rangle\Big)}_{\text{energy eigenvalue } 0\hbar\omega}\right]. \tag{3.9.26}$$

Having the machinery set up, we can also ask more complicated questions, such as: If we have $\uparrow_x\uparrow_x$ at the initial time $t = 0$, what is the probability of $\uparrow_x\uparrow_x$ at the later time t? We begin with noting that

$$|\uparrow_x\uparrow_x\rangle = \frac{1}{2}\Big(|\uparrow_z\uparrow_z\rangle + |\uparrow_z\downarrow_z\rangle + |\downarrow_z\uparrow_z\rangle + |\downarrow_z\downarrow_z\rangle\Big), \tag{3.9.27}$$

so that

$$\psi(0) = \frac{1}{2}\begin{pmatrix} 1 \\ 1 \\ 1 \\ 1 \end{pmatrix} \tag{3.9.28}$$

and

$$\psi(t) = \mathrm{e}^{-\mathrm{i}Ht/\hbar}\psi(0) = \frac{1}{2}\begin{pmatrix} \mathrm{e}^{\mathrm{i}\omega t} \\ \mathrm{e}^{-2\mathrm{i}\omega t} \\ \mathrm{e}^{-2\mathrm{i}\omega t} \\ \mathrm{e}^{\mathrm{i}\omega t} \end{pmatrix}. \tag{3.9.29}$$

The probability amplitude $\langle \uparrow_x\uparrow_x, t | \uparrow_x\uparrow_x, 0\rangle$ is then

$$\underbrace{\frac{1}{2}(1, 1, 1, 1)}_{\langle \uparrow_x\uparrow_x, t| \,\widehat{=}} \underbrace{\psi(t)}_{\widehat{=}\,|\uparrow_x\uparrow_x, 0\rangle} = \frac{1}{2}(\mathrm{e}^{\mathrm{i}\omega t} + \mathrm{e}^{-2\mathrm{i}\omega t}) = \mathrm{e}^{-\frac{\mathrm{i}}{2}\omega t}\cos\left(\frac{3}{2}\omega t\right), \tag{3.9.30}$$

so that the probability

$$\mathrm{prob}(\uparrow_x\uparrow_x \text{ at } t = 0 \ \to \ \uparrow_x\uparrow_x \text{ at } t) = \cos\left(\frac{3}{2}\omega t\right)^2$$

$$= \frac{1}{2} + \frac{1}{2}\cos(3\omega t) \tag{3.9.31}$$

oscillates with frequency 3ω. This is as it should be, because the decomposition

$$|\uparrow_x \uparrow_x\rangle = \frac{1}{\sqrt{2}}\left[\underbrace{\left(|\uparrow_z \uparrow_z\rangle + |\downarrow_z \downarrow_z\rangle\right)/\sqrt{2}}_{\text{energy} = -\hbar\omega} + \underbrace{\left(|\uparrow_z \downarrow_z\rangle + |\downarrow_z \uparrow_z\rangle\right)/\sqrt{2}}_{\text{energy} = 2\hbar\omega}\right] \quad (3.9.32)$$

shows that eigenstates of H with energies $-\hbar\omega$ and $2\hbar\omega$ are involved, so that

$$\left[2\hbar\omega - (-\hbar\omega)\right]/\hbar = 3\omega, \quad \text{indeed.} \quad\quad (3.9.33)$$

3-11 What is the probability of finding $\uparrow_y \uparrow_y$ at time t, if we have $\uparrow_y \uparrow_y$ at $t = 0$?

3-12 Same question for $\uparrow_x \downarrow_x$ at $t = 0$ and later.

Chapter 4

Motion along the x Axis

4.1 Kets, bras, wave functions

On page 70, at the beginning of Section 2.20, we noted the possibility of measurements that have a continuum of possible measurement results, rather than a finite number of outcomes. The prime example is the position of an atom along a line, the x axis, for instance.

We look for the atom and find it at x, but not at x'. Following the structure established for discrete observables A and their eigenkets $|a_k\rangle$ to measurement result a_k, we now introduce ket $|x\rangle$ for "atom at position x", and its bra $\langle x| = |x\rangle^\dagger$, and note

$$\langle x|x'\rangle = 0 \quad \text{if} \quad x \neq x', \tag{4.1.1}$$

saying: "an atom at x' is not at x", but do not specify right now the right-hand side for $x = x'$.

The analog of

$$|\ \rangle = \sum_k |a_k\rangle\langle a_k|\ \rangle = \sum_k |a_k\rangle\psi_k \tag{4.1.2}$$

with the *probability amplitudes* $\psi_k = \langle a_k|\ \rangle$ is then

$$|\ \rangle = \int \mathrm{d}x\, |x\rangle\langle x|\ \rangle = \int \mathrm{d}x\, |x\rangle\psi(x) \tag{4.1.3}$$

with the *wave function* $\psi(x) = \langle x|\ \rangle$, where, fitting to the continuous character of x, we perform an integration as the continuum analog of a summation. Then, as the analogs of

$$\langle 1|2\rangle = \sum_k \langle 1|a_k\rangle\langle a_k|2\rangle = \sum_k \psi_k^{(1)*}\psi_k^{(2)}\,, \qquad (4.1.4)$$

we have

$$\langle 1|2\rangle = \int dx\,\langle 1|x\rangle\langle x|2\rangle = \int dx\,\psi^{(1)}(x)^*\psi^{(2)}(x) \qquad (4.1.5)$$

or

$$\langle 1|2\rangle = \langle 1|\left(\int dx\,|x\rangle\langle x|\right)|2\rangle\,. \qquad (4.1.6)$$

Accordingly, the *completeness* of the $|x\rangle$ kets and the $\langle x|$ bras is expressed by

$$\int dx\,|x\rangle\langle x| = 1\,, \qquad (4.1.7)$$

the analog of

$$\sum_k |a_k\rangle\langle a_k| = 1\,. \qquad (4.1.8)$$

The square of the identity is the identity itself, $1^2 = 1$. In the discrete case this translates into

$$\sum_j |a_j\rangle\langle a_j| \sum_k |a_k\rangle\langle a_k| = \sum_j |a_j\rangle\langle a_j|\,, \qquad (4.1.9)$$

which is a familiar consequence of the orthonormality of the kets $|a_k\rangle$,

$$\langle a_j|a_k\rangle = \delta_{jk}\,. \qquad (4.1.10)$$

For the continuum of $|x\rangle$ kets, we need the appropriate generalization of δ_{jk}, the discrete Kronecker δ symbol, the right-hand side of

$$\langle x|x'\rangle = ? \qquad (4.1.11)$$

to ensure that $1^2 = 1$ translates into

$$\int dx\,|x\rangle\langle x| \int dx'\,|x'\rangle\langle x'| = \int dx\,|x\rangle \underbrace{\int dx'\,\langle x|x'\rangle\langle x'|}_{=\,\langle x|\ \text{(needed)}}$$

$$= \int dx\,|x\rangle\langle x|\,. \qquad (4.1.12)$$

Following Paul A. M. Dirac, we write

$$\langle x|x'\rangle = \delta(x - x'), \tag{4.1.13}$$

which states the orthonormality of the continuum of position states in close analogy to the discrete statement (4.1.10). On the right-hand side we have the so-called *Dirac δ function*, whose defining property is

$$\int dx'\, \delta(x - x')f(x') = f(x) \tag{4.1.14}$$

for all continuous functions $f(x)$. Clearly, this is the continuum analog of

$$\sum_k \delta_{jk} f_k = f_j \tag{4.1.15}$$

in all respects, except for one: whereas the values of δ_{jk} form an ordinary row for each j, and an ordinary column for each k, the Dirac δ function is no ordinary function of x' for any fixed value of x, because it would have to possess extraordinary, yes contradictory, properties: We would need $\delta(x - x') = 0$ if $x' \neq x$ and at the same time

$$\int dx'\, \delta(x - x') = 1, \tag{4.1.16}$$

that is: unit area below a curve that coincides with the x' axis everywhere except at $x' = x$. No ordinary function can be like this.

In fact, the Dirac δ function is not a function in the usual sense — mathematically speaking it is a distribution — but rather an integral kernel that implements the mapping

function $f \longrightarrow$ its value $f(x)$ at x,

$$f \longrightarrow \int dx'\, \delta(x - x')f(x') = f(x). \tag{4.1.17}$$

For this to be a well defined procedure, function f must have a well defined value at x, which is to say that it should be continuous there. Otherwise it could matter whether we approach x from the left or the right.

It is often expedient to regard the Dirac δ function as the limit of ordinary functions that are very strongly peaked at the distinguished point.

The general recipe is:

(1) Take a function $D(\phi)$ with the properties

   (a) $D(\phi) \to 0$ as $|\phi| \to \infty$,
   (b) $D(\phi) \to 1$ as $\phi \to 0$;

(2) define, for $\epsilon > 0$,

$$\delta(x - x'; \epsilon) = \int \frac{dk}{2\pi} D(\epsilon k)\, e^{ik(x - x')} \,;$$

(4.1.18)

(3) then

$$\delta(x - x') = \lim_{\epsilon \to 0} \delta(x - x'; \epsilon) \,;$$

in the sense of

$$f(x) = \lim_{\epsilon \to 0} \int dx'\, \delta(x - x'; \epsilon) f(x') \,.$$

(4.1.19)

We say that such an ordinary function $\delta(x - x'; \epsilon)$ is a *model* for the Dirac δ function.

A simple example is

$$D(\phi) = e^{-|\phi|} \,,$$

(4.1.20)

for which

$$\begin{aligned}
\delta(x - x'; \epsilon) &= \int \frac{dk}{2\pi}\, e^{-|\epsilon k|}\, e^{ik(x - x')} \\
&= \frac{1}{\pi} \mathrm{Re}\left(\int_0^\infty dk\, e^{-\epsilon k + ik(x - x')} \right) \\
&= \frac{1}{\pi} \mathrm{Re}\left(\frac{1}{\epsilon - i(x - x')} \right) \\
&= \frac{1}{\pi} \frac{\epsilon}{\epsilon^2 + (x - x')^2} \,.
\end{aligned}$$

(4.1.21)

Its rough graphical representation is typical:

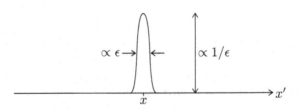

a very narrow, very high peak centered at $x = x'$. The area under the function is

$$\int dx'\, \delta(x - x'; \epsilon) = \frac{1}{\pi} \int dx'\, \frac{\epsilon}{\epsilon^2 + (x - x')^2}$$

$$= \left[\frac{1}{\pi} \arctan \frac{x - x'}{\epsilon} \right]_{x'=-\infty}^{\infty}$$

$$= \frac{1}{\pi} \left(\frac{\pi}{2} - (-\frac{\pi}{2}) \right) = 1\,, \qquad (4.1.22)$$

as it should be. So, any continuous function $f(x')$ that changes little over the distance where $\delta(x - x'; \epsilon)$ is markedly different from 0,

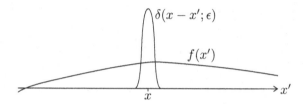

can be replaced by its value at x without affecting the value of the integral:

$$\int dx'\, \delta(x - x'; \epsilon) \underbrace{f(x')}_{\cong\, f(x)} \cong f(x) \underbrace{\int dx'\, \delta(x - x'; \epsilon)}_{=\,1}\,, \qquad (4.1.23)$$

and thus we have a model for the Dirac δ function, indeed.

4-1 Another important example is $D(\phi) = e^{-\frac{1}{2}\phi^2}$. Evaluate $\delta(x - x'; \epsilon)$ for this case and discuss its properties. The gaussian integral

$$\int dx\, e^{-ax^2 + 2bx} = \sqrt{\frac{\pi}{a}}\, e^{b^2/a} \quad \text{for} \quad \text{Re}(a) \geq 0\,, \qquad (4.1.24)$$

which is named after Karl F. Gauss, will be useful.

One more example is the subject matter of Exercise 2-8 on page 56 in *Perturbed Evolution*.

The observation that the orthonormality of the $|x\rangle$ kets is expressed by

$$\langle x|x'\rangle = \delta(x - x') = \begin{cases} 0 & \text{if } x \neq x' \\ \infty & \text{if } x = x' \end{cases} \quad \text{(in some sense)} \qquad (4.1.25)$$

reminds us of the overidealization inherent in identifying $|x\rangle$ with "atom at position x". It overidealizes the realistic physical situation in which an absolutely precise position can never be known, because such absolute precision is an illusion, resources are always limited. You can locate an atom with any desired finite precision (given large finite resources), but not with infinite precision. In short, the kets $|x\rangle$ do not refer to any real physical state of the atom, but nevertheless they are extremely useful mathematical entities, without which we would not be able to go about continuous degrees of freedom efficiently.

4.2 Position operator

The bra-ket product of (4.1.5),

$$\langle 1|2\rangle = \int dx \, \psi^{(1)}(x)^* \psi^{(2)}(x) \tag{4.2.1}$$

has the normalization statement

$$1 = \langle \, | \, \rangle = \int dx \, \psi(x)^* \psi(x) = \int dx \, |\psi(x)|^2 \tag{4.2.2}$$

as a particular case. It says that the area under the curve $|\psi(x)|^2$ is unity:

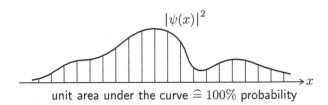

unit area under the curve $\hat{=}$ 100% probability

just stating that there is unit probability of finding the atom *somewhere* when we look for it. Clearly, then, any slice of the area,

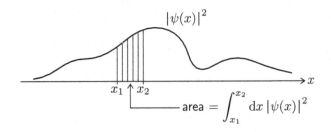

is the probability of finding the atom in this x range:

$$\text{prob}(x_1 < x < x_2) = \int_{x_1}^{x_2} \mathrm{d}x \, |\psi(x)|^2 . \qquad (4.2.3)$$

And so we note that $|\psi(x)|^2$ is a *probability density*, not a probability itself. It turns into a probability upon multiplication with $\mathrm{d}x$,

$$\text{prob}(\text{atom in } x \cdots x + \mathrm{d}x) = \mathrm{d}x |\psi(x)|^2 = \langle \, |x\rangle \mathrm{d}x \langle x| \, \rangle . \qquad (4.2.4)$$

This is the expectation value of $|x\rangle \mathrm{d}x \langle x|$, which therefore has the meaning of "projector on the infinitesimal vicinity of x". It is the infinitesimal version of

$$\int_{x_1}^{x_2} \mathrm{d}x \, |x\rangle\langle x| = \text{projector on the range } x_1 < x < x_2 \qquad (4.2.5)$$

whose expectation value is $\text{prob}(x_1 < x < x_2)$, given in (4.2.3) above.

With these probabilities at hand, we can evaluate weighted averages, *mean values*, of x, x^2, ..., such as

$$\begin{aligned}
\overline{x} &= \int \mathrm{d}x \, x |\psi(x)|^2 = \int \mathrm{d}x \, \langle \, |x\rangle x \langle x| \, \rangle \\
&= \langle \, | \left(\int \mathrm{d}x \, |x\rangle x \langle x| \right) | \, \rangle
\end{aligned} \qquad (4.2.6)$$

and

$$\begin{aligned}
\overline{x^2} &= \int \mathrm{d}x \, x^2 |\psi(x)|^2 = \int \mathrm{d}x \, \langle \, |x\rangle x^2 \langle x| \, \rangle \\
&= \langle \, | \left(\int \mathrm{d}x \, |x\rangle x^2 \langle x| \right) | \, \rangle
\end{aligned} \qquad (4.2.7)$$

and so forth. We thus identify the *position operator*

$$X = \int \mathrm{d}x \, |x\rangle x \langle x| \qquad (4.2.8)$$

whose expectation value is the mean value of x,

$$\langle X \rangle = \int \mathrm{d}x \, x |\psi(x)|^2 . \qquad (4.2.9)$$

This is a consistent identification because the expectation value $\langle X^2 \rangle$ is really the mean value of x^2, inasmuch as

$$
\begin{aligned}
X^2 &= \left(\int \mathrm{d}x\, |x\rangle x \langle x| \right) \left(\int \mathrm{d}x'\, |x'\rangle x' \langle x'| \right) \\
&= \int \mathrm{d}x\, |x\rangle x \underbrace{\int \mathrm{d}x'\, \delta(x - x') x' \langle x'|}_{= x\langle x|} \\
&= \int \mathrm{d}x\, |x\rangle x^2 \langle x| \,,
\end{aligned}
\tag{4.2.10}
$$

as it should be. These generalize immediately to arbitrary functions of X,

$$
f(X) = \int \mathrm{d}x\, |x\rangle f(x) \langle x| \,,
\tag{4.2.11}
$$

which we recognize as the continuous analog of the discrete construction

$$
f(A) = \sum_k |a_k\rangle f(a_k) \langle a_k|
\tag{4.2.12}
$$

of (2.20.15).

The eigenvalue equation

$$
f(A)|a_k\rangle = |a_k\rangle f(a_k)
\tag{4.2.13}
$$

has its obvious analog as well,

$$
f(X)|x\rangle = |x\rangle f(x) \,.
\tag{4.2.14}
$$

Let us check this:

$$
f(X)|x\rangle = \int \mathrm{d}x'\, |x'\rangle f(x') \underbrace{\langle x'|x\rangle}_{= \delta(x' - x)} = |x\rangle f(x) \,,
\tag{4.2.15}
$$

indeed. Everything fits together very neatly.

4-2   Find the constant in $\psi(x) = (\mathrm{const})\, e^{-\kappa|x|}$, $\kappa > 0$, that is needed for proper normalization of $\psi(x)$. Then evaluate $\langle X \rangle$ and $\langle X^2 \rangle$.

4-3   More generally, evaluate $\left\langle e^{\mathrm{i}kX} \right\rangle$ for real k, and extract $\langle X^n \rangle$ for $n = 0, 1, 2, \ldots$ by expanding in powers of k.

4.3   Momentum operator

Since the eigenkets of position operator X are labeled by the continuous parameter x, it is natural to ask what happens if we differentiate with respect to x. It is slightly more systematic to look at bras $\langle x|$ first, so we are asking

$$\frac{\partial}{\partial x}\langle x| = ? \tag{4.3.1}$$

The outcome is a linear combination of the bras $\langle x|$, of course, and that is also the case for higher derivatives. We can talk about all of them at once by making use of Brook Taylor's seminal theorem,

$$f(x + x') = \sum_{k=0}^{\infty} \frac{1}{k!} x'^k \left(\frac{\partial}{\partial x}\right)^k f(x), \tag{4.3.2}$$

which we write compactly as

$$f(x + x') = e^{x'\frac{\partial}{\partial x}} f(x). \tag{4.3.3}$$

When applied to $\langle x|$, it reads

$$\langle x + x'| = e^{x'\frac{\partial}{\partial x}} \langle x|. \tag{4.3.4}$$

But the mapping $\langle x| \rightarrow \langle x + x'|$ is unitary,

$$\langle x + x'| = \langle x|U \tag{4.3.5}$$

with

$$U = \int dx \, |x\rangle\langle x|U = \int dx \, |x\rangle\langle x + x'|. \tag{4.3.6}$$

See

$$U^\dagger U = \int dx'' \, |x'' + x'\rangle \underbrace{\langle x''| \int dx \, |x\rangle}\langle x + x'|$$

$$\rightarrow \int dx \, \delta(x'' - x)$$

$$= \int dx'' \, |x'' + x'\rangle\langle x'' + x'|$$

$$= \int dx \, |x\rangle\langle x| = 1 \tag{4.3.7}$$

and

$$UU^\dagger = \int \mathrm{d}x\, |x\rangle \underbrace{\langle x + x'| \int \mathrm{d}x''\, |x'' + x'\rangle\langle x''|}_{\rightarrow \int \mathrm{d}x''\, \delta(x - x'')}$$

$$= \int \mathrm{d}x\, |x\rangle\langle x| = 1\,. \tag{4.3.8}$$

As a unitary operator, U is the exponential of some hermitian operator (times i), and the exponential will be linear in x', so

$$U = \mathrm{e}^{\mathrm{i}x'P/\hbar} \tag{4.3.9}$$

with $P = P^\dagger$ of metrical dimension action/distance = momentum. Indeed, this P is the *momentum operator*, an identification to be justified below. Its action on position eigenstates is stated by

$$\langle x|\, \mathrm{e}^{\mathrm{i}x'P/\hbar} = \langle x + x'| = \mathrm{e}^{x'\frac{\partial}{\partial x}} \langle x|\,. \tag{4.3.10}$$

A term-by-term comparison of an expansion in powers of x' tells us that

$$\langle x|P = \frac{\hbar}{\mathrm{i}} \frac{\partial}{\partial x} \langle x|\,,$$

$$\langle x|P^2 = \left(\frac{\hbar}{\mathrm{i}} \frac{\partial}{\partial x}\right)^2 \langle x|\,,$$

$$\langle x|P^n = \left(\frac{\hbar}{\mathrm{i}} \frac{\partial}{\partial x}\right)^n \langle x|\,, \tag{4.3.11}$$

for $n = 0, 1, 2, \dots$. The fundamental statement is the one for $n = 1$,

$$P \cong \frac{\hbar}{\mathrm{i}} \frac{\partial}{\partial x} \qquad \text{(when acting on position bras)}, \tag{4.3.12}$$

the historical starting point of Erwin Schrödinger's formulation of quantum mechanics in terms of differential equations obeyed by the position wave function $\psi(x)$.

Note, however, the important detail that this identification of the position operator with the differential operator $\dfrac{\hbar}{\mathrm{i}} \dfrac{\partial}{\partial x}$ is only correct when we consider the action of P on bra $\langle x|$. In the case of P acting on ket $|x\rangle$, we have the adjoint statement

$$P|x\rangle = \mathrm{i}\hbar \frac{\partial}{\partial x} |x\rangle \tag{4.3.13}$$

and the identification would read

$$P \cong i\hbar \frac{\partial}{\partial x} \qquad \text{(when acting on position kets)} \qquad (4.3.14)$$

and, clearly, these are just two of many identifications of this sort.

4.4 Heisenberg's commutation relation

Now consider the products XP and PX. We have

$$\langle x|XP = x\langle x|P = x\frac{\hbar}{i}\frac{\partial}{\partial x}\langle x| \qquad (4.4.1)$$

and

$$\langle x|PX = \frac{\hbar}{i}\frac{\partial}{\partial x}\langle x|X = \frac{\hbar}{i}\frac{\partial}{\partial x}(x\langle x|) = x\frac{\hbar}{i}\frac{\partial}{\partial x}\langle x| + \frac{\hbar}{i}\langle x| \qquad (4.4.2)$$

so that

$$\langle x|(XP - PX) = i\hbar\langle x| \qquad (4.4.3)$$

for all position bras $\langle x|$. But they are complete, so that

$$XP - PX = i\hbar \qquad (4.4.4)$$

or

$$[X,P] = i\hbar. \qquad (4.4.5)$$

This is the *Heisenberg commutator*, the famous position-momentum commutation relation, discovered by Werner Heisenberg (with some help from Max Born), the starting point of Heisenberg's formulation of quantum mechanics in terms of operators and their evolution.

4.5 Position-momentum transformation function

The Heisenberg commutator is invariant under the exchange of X and P, in the sense of

$$X \to P, \qquad P \to -X \qquad (4.5.1)$$

since

$$[X,P] \to [P,-X] = [X,P], \qquad (4.5.2)$$

which tells us that the properties of operator P are largely the same as those of operator X, their roles are interchangeable. As an immediate consequence, there must also be a continuum of eigenbras and eigenkets of P,

$$\langle p|P = p\langle p|\,, \quad P|p\rangle = |p\rangle p \tag{4.5.3}$$

that are orthonormal (in the Dirac δ function sense)

$$\langle p|p'\rangle = \delta(p - p') \tag{4.5.4}$$

and complete

$$\int \mathrm{d}p\, |p\rangle\langle p| = 1\,. \tag{4.5.5}$$

To relate them to the position eigenstates

$$|p\rangle = \int \mathrm{d}x\, |x\rangle\langle x|p\rangle \tag{4.5.6}$$

we need the *transformation function* $\langle x|p\rangle$. We find it by exploiting the eigenket property,

$$\langle x|P|p\rangle = \langle x|p\rangle p\,, \tag{4.5.7}$$

and Schrödinger's differential operator identification,

$$\langle x|P|p\rangle = \frac{\hbar}{\mathrm{i}}\frac{\partial}{\partial x}\langle x|p\rangle\,, \tag{4.5.8}$$

for first establishing the differential equation

$$\frac{\partial}{\partial x}\langle x|p\rangle = \frac{\mathrm{i}}{\hbar}p\langle x|p\rangle\,, \tag{4.5.9}$$

and then solving it,

$$\langle x|p\rangle = C(p)\,\mathrm{e}^{\mathrm{i}xp/\hbar}\,, \tag{4.5.10}$$

where $C(p)$ is, for now, an arbitrary function of p.

We learn something about it upon combining the orthonormality relation of the x states with the completeness relation for the p states,

$$\delta(x - x') = \langle x|x' \rangle = \int dp \, \langle x|p \rangle \langle p|x' \rangle$$

$$= \int dp \, |C(p)|^2 \, e^{ip(x - x')/\hbar}$$

$$= \int \frac{dk}{2\pi} 2\pi\hbar |C(p)|^2 \, e^{ik(x - x')} \,, \qquad (4.5.11)$$

where $k = p/\hbar$. Compare this with

$$\delta(x - x') = \lim_{\epsilon \to 0} \delta(x - x'; \epsilon)$$

$$= \lim_{\epsilon \to 0} \int \frac{dk}{2\pi} D(\epsilon k) \, e^{ik(x - x')} \qquad (4.5.12)$$

in (4.1.18), and conclude that consistency requires

$$2\pi\hbar |C(p)|^2 = \lim_{\epsilon \to 0} D(\epsilon k) = 1 \qquad (4.5.13)$$

or

$$|C(p)| = \frac{1}{\sqrt{2\pi\hbar}} \,. \qquad (4.5.14)$$

It is up to us to assign arbitrary p dependent phases to $|\ \rangle = |p\rangle$, but there is absolutely no benefit in choosing anything else but the trivial phase for the kets $|p\rangle$. Then

$$\langle x|p \rangle = \frac{1}{\sqrt{2\pi\hbar}} \, e^{ixp/\hbar} \,. \qquad (4.5.15)$$

Reading this as the wave function $\psi(x) = \langle x| \ \rangle$ to the ket $|\ \rangle = |p\rangle$, we note that a momentum eigenstate is a (plane) wave with wavelength

$$\lambda = \frac{2\pi\hbar}{p} \qquad \text{or} \qquad \lambda = \frac{\hbar}{p} \qquad (4.5.16)$$

since an increment of x by λ changes $xp/\hbar$ by 2π and thus corresponds to one full period of this wave.

The relation $\lambda = \hbar/p$ is known as the de Broglie relation. It was Prince Louis-Victor de Broglie's work that triggered Erwin Schrödinger's interest, because once you have such waves, you start wondering about the underlying wave equation — and that is what became known as the Schrödinger equation; more about it shortly.

4-4 Atoms in a gas at room temperature have an average kinetic energy $p^2/(2M)$ (M = mass of the atom) of about $0.025\,\mathrm{eV}$. What would be the resulting de Broglie wavelength for silver atoms? For helium atoms?

4.6 Expectation values

By essentially reversing the above argument, we can establish the action of position operator X on momentum kets $|p\rangle$,

$$
\begin{aligned}
\langle x|X|p\rangle &= x\langle x|p\rangle = x\frac{1}{\sqrt{2\pi\hbar}}\,\mathrm{e}^{\mathrm{i}xp/\hbar} \\
&= \frac{\hbar}{\mathrm{i}}\frac{\partial}{\partial p}\langle x|p\rangle = \langle x|\frac{\hbar}{\mathrm{i}}\frac{\partial}{\partial p}|p\rangle\,,
\end{aligned}
\tag{4.6.1}
$$

so that

$$
X|p\rangle = \frac{\hbar}{\mathrm{i}}\frac{\partial}{\partial p}|p\rangle
\tag{4.6.2}
$$

follows upon invoking the completeness of the x states once more.

Just as we did for operator X, we can speak about functions of operator P in the usual way, beginning with

$$
\begin{aligned}
P &= \int \mathrm{d}p\,|p\rangle p\langle p|\,, \\
P^2 &= \int \mathrm{d}p\,|p\rangle p^2\langle p|\,, \\
&\;\;\vdots \\
P^n &= \int \mathrm{d}p\,|p\rangle p^n\langle p|\,,
\end{aligned}
\tag{4.6.3}
$$

and finally, most generally,

$$
f(P) = \int \mathrm{d}p\,|p\rangle f(p)\langle p|\,,
\tag{4.6.4}
$$

the obvious analog of the position version (4.2.11),

$$
f(X) = \int \mathrm{d}x\,|x\rangle f(x)\langle x|\,,
\tag{4.6.5}
$$

and the earlier discrete version of (2.20.15),

$$f(A) = \sum_{k=1}^{n} |a_k\rangle f(a_k)\langle a_k| \,. \tag{4.6.6}$$

Accordingly, we evaluate expectation values, or average values, of such functions by integration,

$$\langle f(X) \rangle = \int dx \, |\psi(x)|^2 f(x) \,,$$

$$\langle g(P) \rangle = \int dp \, |\psi(p)|^2 g(p) \,, \tag{4.6.7}$$

where

$$\psi(x) = \langle x| \, \rangle \tag{4.6.8}$$

is the position wave function, and

$$\psi(p) = \langle p| \, \rangle \tag{4.6.9}$$

is the momentum wave function. The transformation functions

$$\langle x|p \rangle = \frac{1}{\sqrt{2\pi\hbar}} \, e^{ixp/\hbar} \,,$$

$$\langle p|x \rangle = \frac{1}{\sqrt{2\pi\hbar}} \, e^{-ipx/\hbar} \tag{4.6.10}$$

relate them to each other,

$$\psi(x) = \int dp \, \langle x|p\rangle\langle p| \, \rangle = \int dp \, \frac{e^{ixp/\hbar}}{\sqrt{2\pi\hbar}} \psi(p) \,,$$

$$\psi(p) = \int dx \, \langle p|x\rangle\langle x| \, \rangle = \int dx \, \frac{e^{-ipx/\hbar}}{\sqrt{2\pi\hbar}} \psi(x) \,. \tag{4.6.11}$$

Mathematically speaking, these are Fourier transformations and the pair of relations states Fourier's theorem about his transformation and its inverse.

We can do, what Jean B. J. Fourier could not, namely express his theorem compactly, with the aid of Dirac's δ function. For, if we consider the iteration of (4.6.11),

$$\psi(x) = \int dp \, \frac{e^{ixp/\hbar}}{\sqrt{2\pi\hbar}} \int dx' \, \frac{e^{-ipx'/\hbar}}{\sqrt{2\pi\hbar}} \psi(x')$$

$$= \int dx' \left(\int dp \, \frac{e^{ip(x - x')/\hbar}}{2\pi\hbar} \right) \psi(x') \tag{4.6.12}$$

we read off that

$$\int \mathrm{d}p \, \frac{e^{\mathrm{i}p(x - x')/\hbar}}{2\pi\hbar} = \delta(x - x') , \tag{4.6.13}$$

the Fourier representation of the δ function — that we have implicitly used already in (4.5.11) with (4.5.14) — which is in fact the basis of the construction (4.1.18).

Given that

$$\langle x|P| \, \rangle = \frac{\hbar}{\mathrm{i}} \frac{\partial}{\partial x} \langle x| \, \rangle = \frac{\hbar}{\mathrm{i}} \frac{\partial}{\partial x} \psi(x) \tag{4.6.14}$$

we can evaluate the expectation value of P in terms of the x wave function, either as

$$\langle P \rangle = \int \mathrm{d}x \, \langle \, |x\rangle\langle x|P| \, \rangle$$
$$= \int \mathrm{d}x \, \psi(x)^* \frac{\hbar}{\mathrm{i}} \frac{\partial}{\partial x} \psi(x) \tag{4.6.15}$$

or as

$$\langle P \rangle = \int \mathrm{d}x \, \langle \, |P|x\rangle\langle x| \, \rangle$$
$$= \int \mathrm{d}x \, \left(\mathrm{i}\hbar \frac{\partial}{\partial x} \psi(x)^* \right) \psi(x) . \tag{4.6.16}$$

These must, of course, give the same result, and indeed they do because their difference

$$\int \mathrm{d}x \left[\psi(x)^* \frac{\hbar}{\mathrm{i}} \frac{\partial}{\partial x} \psi(x) - \left(\mathrm{i}\hbar \frac{\partial}{\partial x} \psi(x)^* \right) \psi(x) \right]$$
$$= \frac{\hbar}{\mathrm{i}} \int \mathrm{d}x \, \frac{\partial}{\partial x} [\psi(x)^* \psi(x)]$$
$$= \left[\frac{\hbar}{\mathrm{i}} |\psi(x)|^2 \right]_{x=-\infty}^{\infty} = 0 \tag{4.6.17}$$

vanishes since $|\psi(x)|^2 \to 0$ for $x \to \pm\infty$, as is ensured by

$$\int \mathrm{d}x \, |\psi(x)|^2 = 1 . \tag{4.6.18}$$

4-5 The last argument is, in fact, a bit sloppy. Show this by giving an example of a function $\psi(x)$, for which (4.6.18) holds, but $|\psi(x)|^2 \to 0$ for $x \to \pm\infty$ is not true. How, then, can you ensure the null result of (4.6.17)?

4-6 Show that the following expressions are equivalent:

$$\left\langle P^2 \right\rangle = -\hbar^2 \int dx\, \psi(x)^* \left(\frac{\partial}{\partial x}\right)^2 \psi(x)$$

$$= -\hbar^2 \int dx\, \psi(x) \left(\frac{\partial}{\partial x}\right)^2 \psi(x)^*$$

$$= \hbar^2 \int dx \left|\frac{\partial}{\partial x}\psi(x)\right|^2 .$$

Should you be bothered by the lesson of Exercise 4-5?

4-7 For the $\psi(x)$ in Exercise 4-2 on page 116, find $\psi(p)$, then calculate $\langle P \rangle$ and $\langle P^2 \rangle$ in at least two different ways each.

4.7 Uncertainty relation

For any hermitian A, be it a function of X, or of P, or of both (or, possibly, involving further degrees of freedom) we have its expectation value $\langle A \rangle$, so that the difference $A - \langle A \rangle$ is an operator with vanishing expectation value. A measurement of $A - \langle A \rangle$ gives results that are positive or negative, such that their weighted sum, weighted by the relative frequency of occurrence, is zero. As a mathematical statement,

$$\langle (A - \langle A \rangle) \rangle = 0 , \tag{4.7.1}$$

it borders on a triviality. Yet, how this zero average comes about, can be very different. There can be lots of measurement results of A close to $\langle A \rangle$ and scattered about this value, or lots of results quite different from $\langle A \rangle$, some much larger, some much smaller. To get a quantitative handle on such differences in the data scatter, we evaluate the expectation value of $(A - \langle A \rangle)^2$. It is the mean value of positive numbers (or, at least, not negative ones) and is therefore itself positive. This permits to identify its square root, δA, with the *spread* of the measurement results of $A - \langle A \rangle$, or in fact the spread of A itself. Thus we write

$$\delta A = \sqrt{\left\langle (A - \langle A \rangle)^2 \right\rangle} \tag{4.7.2}$$

as the technical definition of δA. In view of the identity

$$\left\langle (A - \langle A\rangle)^2 \right\rangle = \left\langle \left(A^2 - 2\langle A\rangle A + \langle A\rangle^2 \right) \right\rangle$$
$$= \left\langle A^2 \right\rangle - 2\langle A\rangle\langle A\rangle + \langle A\rangle^2$$
$$= \left\langle A^2 \right\rangle - \langle A\rangle^2 , \tag{4.7.3}$$

we note that

$$\delta A = \sqrt{\left\langle A^2 \right\rangle - \langle A\rangle^2} \tag{4.7.4}$$

is an equivalent definition.

Now, consider two observables specified by operators A and B, along with their expectation values, $\langle A\rangle, \langle B\rangle$ and their spreads $\delta A, \delta B$. Both A and B are, of course, meant to be hermitian in the present context — otherwise $(A - \langle A\rangle)^2$ would not be assuredly positive — and so the auxiliary operator

$$C = \delta B\, (A - \langle A\rangle) + \mathrm{i}\delta A\, (B - \langle B\rangle) \tag{4.7.5}$$

is not hermitian, it differs from its adjoint

$$C^\dagger = \delta B\, (A - \langle A\rangle) - \mathrm{i}\delta A\, (B - \langle B\rangle) \tag{4.7.6}$$

by a change of sign for the second summand. Yet, the products

$$C^\dagger C \quad \text{and} \quad CC^\dagger \tag{4.7.7}$$

are both positive operators in the sense that their expectation values,

$$\left\langle C^\dagger C \right\rangle = \mathrm{tr}\{\rho C^\dagger C\} \geq 0 ,$$
$$\left\langle CC^\dagger \right\rangle = \mathrm{tr}\{\rho CC^\dagger\} \geq 0 , \tag{4.7.8}$$

are positive for *all* statistical operators ρ. This is easy to see as soon as we represent ρ as one of its many blends,

$$\rho = \sum_k |k\rangle w_k \langle k| , \qquad w_k > 0 , \qquad \sum_k w_k = 1 , \tag{4.7.9}$$

where the kets $|k\rangle$ are normalized, $\langle k|k\rangle = 1$, but otherwise arbitrary, and the weights w_k are only restricted by the stated positivity and normalization to unit sum. Then, for example,

$$\left\langle C^\dagger C \right\rangle = \sum_k \underbrace{w_k}_{>0} \underbrace{\langle k|C^\dagger C|k\rangle}_{= (\langle k|C^\dagger)(C|k\rangle) = (C|k\rangle)^\dagger (C|k\rangle) \geq 0} \geq 0 \tag{4.7.10}$$

indeed.

Writing it out,

$$\begin{aligned}
C^\dagger C &= (\delta B)^2 (A - \langle A\rangle)^2 + (\delta A)^2 (B - \langle B\rangle)^2 \\
&\quad + \mathrm{i}\delta A\,\delta B\Big((A - \langle A\rangle)(B - \langle B\rangle) - (B - \langle B\rangle)(A - \langle A\rangle)\Big) \\
&= (\delta B)^2 (A - \langle A\rangle)^2 + (\delta A)^2 (B - \langle B\rangle)^2 \\
&\quad + \mathrm{i}\delta A\,\delta B\,[A, B]\,,
\end{aligned} \tag{4.7.11}$$

we have

$$0 \leq \left\langle C^\dagger C \right\rangle = 2(\delta A\,\delta B)^2 + \mathrm{i}\delta A\,\delta B\,\langle [A, B]\rangle \tag{4.7.12}$$

and likewise

$$0 \leq \left\langle C C^\dagger \right\rangle = 2(\delta A\,\delta B)^2 - \mathrm{i}\delta A\,\delta B\,\langle [A, B]\rangle\,, \tag{4.7.13}$$

or

$$\delta A\,\delta B \geq -\frac{1}{2}\langle \mathrm{i}[A, B]\rangle \quad \text{and} \quad \delta A\,\delta B \geq +\frac{1}{2}\langle \mathrm{i}[A, B]\rangle\,. \tag{4.7.14}$$

Accordingly,

$$\delta A\,\delta B \geq \frac{1}{2}\big|\langle \mathrm{i}[A, B]\rangle\big|\,. \tag{4.7.15}$$

This is Howard P. Robertson's general version of Werner Heisenberg's uncertainty relation. The original version is obtained by specializing to $A = X, B = P$, for which

$$\mathrm{i}[A, B] = \mathrm{i}\underbrace{[X, P]}_{=\,\mathrm{i}\hbar} = -\hbar\,, \tag{4.7.16}$$

so that

$$\delta X\,\delta P \geq \frac{1}{2}\hbar\,. \tag{4.7.17}$$

This is the celebrated Heisenberg's *uncertainty relation*, perhaps the most famous of all the famous easily-stated results of quantum mechanics. It is sometimes, and a bit misleadingly, referred to as the "uncertainty principle" although it is not a physical principle of any kind, it is a relation, an inequality.

4-8 In the derivation of the general uncertainty relation, we divide by $\delta A\,\delta B$ at one point, which is only permissible if both spreads are truly positive. Why does this restriction not matter in the end? That is: Show that the uncertainty relation applies also when $\delta A = 0$, or $\delta B = 0$, or $\delta A = 0$ and $\delta B = 0$.

What is the physical content of Heisenberg's uncertainty relation for position and momentum? Let us first consider a situation in which the momentum spread δP is very small. Then the momentum is well defined, we can predict with virtual certainty the outcome of a momentum measurement. But this also means that the de Broglie wavelength $\lambda_{\mathrm{dB}} = 2\pi\hbar/p$ of (4.5.16) is a very well defined quantity. Now, to determine the wavelength of any wave with high precision, you need a very long wave train, one that displays many regular oscillations. Such a wave train is spread out over a large area and thus there is no precision at all when you assign a position to this wave train. Accordingly, the outcome of a position measurement will be utterly unpredictable, so that δX has to be large if δP is small.

The other extreme of a small δX, that is of a well defined position, in the sense that we can predict the outcome of a position measurement with great confidence, would in term require a superposition of de Broglie waves with a very wide spread in wavelength, and thus momentum, in order to ensure destructive interference almost everywhere, except for the small x region where the atom is located. Thus a small δX implies a large δP, indeed.

4-9 What is $\delta X\delta P$ for the state specified by the wave function $\psi(x)$ in Exercise 4-2 on page 116?

4-10 In the Stern–Gerlach experiment on page 11, assume that the silver atoms have a speed of $500\,\mathrm{m/s}$ and are collimated by two circular apertures of $1\,\mathrm{mm}$ diameter that are $50\,\mathrm{cm}$ apart. Estimate δX and δP, and then $\delta X\,\delta P/\hbar$.

4-11 For small displacements x, the squared expectation value of the unitary displacement operator $e^{\mathrm{i}xP/\hbar}$ is of the form

$$\left|\left\langle e^{\mathrm{i}xP/\hbar}\right\rangle\right|^2 = 1 - \left(\frac{x}{L}\right)^2 + O(x^4),$$

where L is the so-called *coherence length*. How is L related to the momentum spread δP? How does L compare with the position spread δX?

4.8   State of minimum uncertainty

The derivation of the general uncertainty relation (4.7.15) shows that the equal sign can only hold for states described by kets $|\ \rangle$ that are either eigenkets of C, or of $C^\dagger$, with eigenvalue zero,

$$\text{either} \quad C|\ \rangle = 0 \quad \text{or} \quad C^\dagger|\ \rangle = 0. \tag{4.8.1}$$

Let us see if we can find such a minimum-uncertainty state for $A = X$ and $B = P$. It is clearly enough to consider the case $\langle X \rangle = 0$, $\langle P \rangle = 0$, and so we are looking for a ket that obeys

$$(\delta P\, X + \mathrm{i}\delta X\, P)|\ \rangle = 0 \tag{4.8.2}$$

with $\delta X \delta P = \frac{1}{2}\hbar$; for "$-\mathrm{i}$" rather than "$+\mathrm{i}$" see after (4.8.19) below.

Put bra $\langle x|$ next to this equation to turn it into a differential equation for the wave function $\psi(x) = \langle x|\ \rangle$:

$$0 = \langle x|(\delta P\, X + \mathrm{i}\delta X\, P)|\ \rangle = \Big(\underbrace{\delta P}_{=\,\hbar/(2\delta X)}\, x + \mathrm{i}\delta X \frac{\hbar}{\mathrm{i}} \frac{\partial}{\partial x}\Big) \underbrace{\langle x|\ \rangle}_{=\,\psi(x)} \tag{4.8.3}$$

or

$$\frac{\partial}{\partial x}\psi(x) = -\frac{x}{2(\delta X)^2}\psi(x) \tag{4.8.4}$$

which is solved by

$$\psi(x) = \psi_0\, \mathrm{e}^{-\left(\frac{1}{2}x/\delta X\right)^2} . \tag{4.8.5}$$

The multiplicative constant ψ_0 is fixed partly by normalization,

$$\int \mathrm{d}x\, |\psi(x)|^2 = 1, \tag{4.8.6}$$

and partly by convention, that is: the freedom to choose an arbitrary overall phase factor, which we exploit by insisting on $\psi_0 > 0$. Then the normalization says

$$|\psi_0|^2 \int \mathrm{d}x\, \mathrm{e}^{-\frac{1}{2}(x/\delta X)^2} = 1, \tag{4.8.7}$$

where we meet a gaussian integral, a special case of the integral in (4.1.24), so that

$$|\psi_0|^2 \sqrt{\frac{\pi}{\frac{1}{2}/(\delta X)^2}} = |\psi_0|^2 \sqrt{2\pi}\,\delta X = 1 \tag{4.8.8}$$

or

$$\psi_0 = \frac{(2\pi)^{-\frac{1}{4}}}{\sqrt{\delta X}} \quad \text{for} \quad \psi_0 > 0 \,. \tag{4.8.9}$$

In summary, the quest for a minimum-uncertainty state is answered affirmatively by

$$\psi(x) = \langle x| \ \rangle = \frac{(2\pi)^{-\frac{1}{4}}}{\sqrt{\delta X}} \, e^{-(\frac{1}{2}x/\delta X)^2} \,. \tag{4.8.10}$$

The corresponding momentum wave function is also available,

$$\psi(p) = \langle p| \ \rangle = \int dx \, \langle p|x\rangle\langle x| \ \rangle \tag{4.8.11}$$

or

$$\psi(p) = \frac{1}{\sqrt{2\pi\hbar}} \frac{(2\pi)^{-\frac{1}{4}}}{\sqrt{\delta X}} \int dx \, e^{-(\frac{1}{2}x/\delta X)^2 - ipx/\hbar} \,. \tag{4.8.12}$$

This is another gaussian integral, now with the parameters in (4.1.24) identified as

$$a = \frac{1}{(2\delta X)^2} \,, \qquad b = -\frac{i}{2}\frac{p}{\hbar} \,, \tag{4.8.13}$$

so that

$$\begin{aligned}
\psi(p) &= \frac{1}{\sqrt{2\pi\hbar}} \frac{(2\pi)^{-\frac{1}{4}}}{\sqrt{\delta X}} \sqrt{\pi}\, 2\delta X \, e^{-(p\delta X/\hbar)^2} \\
&= \frac{(2\pi)^{-\frac{1}{4}}}{\sqrt{\frac{\hbar}{2\delta X}}} \, e^{-(p\delta X/\hbar)} \,,
\end{aligned} \tag{4.8.14}$$

or with $\delta P = \dfrac{\hbar}{2}/\delta X$,

$$\psi(p) = \frac{(2\pi)^{-\frac{1}{4}}}{\sqrt{\delta P}} \, e^{-(\frac{1}{2}p/\delta P)^2} \,. \tag{4.8.15}$$

Note how the replacements $x \leftrightarrow p$, $\delta X \leftrightarrow \delta P$ turn $\psi(x)$ and $\psi(p)$ into each other.

It remains to verify, as a check and a protection against silly errors, that the position and momentum uncertainties for these wave functions are indeed equal to δX and δP. Since $\langle X \rangle = 0$ and $\langle P \rangle = 0$ for obvious reasons of symmetry, we are called to verify that

$$\langle X^2 \rangle = (\delta X)^2 \,, \qquad \langle P^2 \rangle = (\delta P)^2 \,, \tag{4.8.16}$$

where it suffices to check one of them. We take $\langle X^2 \rangle$,

$$
\begin{aligned}
\langle X^2 \rangle &= \int \mathrm{d}x\, x^2 |\psi(x)|^2 \\
&= \frac{1}{\sqrt{2\pi}} \frac{1}{\delta X} \int \mathrm{d}x\, x^2\, \mathrm{e}^{-\frac{1}{2}(x/\delta X)^2} .
\end{aligned}
\tag{4.8.17}
$$

This integral is also a gaussian integral of some sort,

$$
\begin{aligned}
\int \mathrm{d}x\, x^2\, \mathrm{e}^{-ax^2} &= -\frac{\partial}{\partial a} \int \mathrm{d}x\, \mathrm{e}^{-ax^2} \\
&= -\frac{\partial}{\partial a} \sqrt{\frac{\pi}{a}} = \frac{1}{2}\sqrt{\frac{\pi}{a^3}} ,
\end{aligned}
\tag{4.8.18}
$$

so that

$$
\begin{aligned}
\langle X^2 \rangle &= \frac{1}{\sqrt{2\pi}} \frac{1}{\delta X} \frac{1}{2} \sqrt{\frac{\pi}{(\frac{1}{2}/(\delta X)^2)^3}} \\
&= \frac{1}{\sqrt{2\pi}} \frac{1}{\delta X} \frac{1}{2} \sqrt{8\pi(\delta X)^6} = (\delta X)^2 ,
\end{aligned}
\tag{4.8.19}
$$

indeed.

Finally, we remark on the apparent option to choose the other sign in (4.8.3). It would lead us first to

$$
\frac{\partial}{\partial x}\psi(x) = \frac{x}{2(\delta X)^2}\psi(x)
\tag{4.8.20}
$$

and then to

$$
\psi(x) = \psi_0\, \mathrm{e}^{(\frac{1}{2}x/\delta X)^2} ,
\tag{4.8.21}
$$

for which there is no way of satisfying the normalization condition

$$
\int \mathrm{d}x\, |\psi(x)|^2 = 1 .
\tag{4.8.22}
$$

As a consequence, this other sign choice is not an actual option.

4-12   Use the three x integrals in Exercise 4-6 on page 125 to calculate $(\delta P)^2$ from $\psi(x)$ and verify thus once more that $\delta X\, \delta P = \frac{1}{2}\hbar$ here.

4.9　Time dependence

So far, all remarks about position X and momentum P referred to one single instant in time. But, of course, atoms move, and so X is naturally dependent on time or, put differently, we can measure the atom's position at various instants and need position operators for all times, $X(t)$. Likewise the momentum operator depends on time, $P(t)$.

Then, also their eigenkets and eigenbras acquire a time dependence, as illustrated by the eigenvalue equation

$$\langle x,t|\, X(t) = x \langle x,t|\,, \tag{4.9.1}$$

which is an equation of the same kind as (3.1.3). Small time increments δt are generated by the Hamilton operator H, small position increments δx by the momentum operator P, so that a change of both gives

$$\delta\langle x,t| = \frac{\mathrm{i}}{\hbar}\langle x,t|\big(P(t)\delta x - H(t)\delta t\big)\,, \tag{4.9.2}$$

which encompasses the Schrödinger equation

$$\mathrm{i}\hbar\frac{\partial}{\partial t}\langle\ldots,t| = \langle\ldots,t|H(t) \tag{4.9.3}$$

as well as

$$\frac{\hbar}{\mathrm{i}}\frac{\partial}{\partial x}\langle x,\ldots| = \langle x,\ldots|P\,, \tag{4.9.4}$$

where the ellipses stand for other, presently irrelevant labels.

The relative sign in

$$P\delta x - H\delta t \tag{4.9.5}$$

originates in the sign convention in (4.3.10), where we wrote $e^{\mathrm{i}x'P/\hbar}$ rather than $e^{-\mathrm{i}x'P/\hbar}$, a choice that was not given any justification then. We shall close this gap now, with a little excursion into classical mechanics.

4.10　Excursion into classical mechanics

The question is, of course, not about the chosen sign of i in all our equations, we could just as well replace i by $-$i everywhere without the slightest consequence (in the early days of quantum mechanics there was a temporary coexistence of both sign conventions, but nowadays everybody uses the same convention, the one that has $[X,P] = \mathrm{i}\hbar$, rather than $-\mathrm{i}\hbar$). The question is about the relative minus sign between $P\delta x$ and $H\delta t$. And that comes straight from classical mechanics, and stays with us in quantum mechanics if we wish, as we do, to maintain as good a resemblance as we can. After all, we borrow classical terminology, such as "Hamilton function" and "momentum" from classical mechanics and should, therefore, do this borrowing consistently.

We recall Joseph L. de Lagrange's variational principle. It says that the actual trajectory is the one for which the action

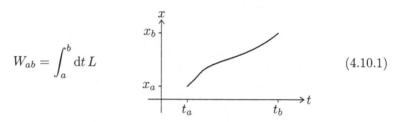

$$W_{ab} = \int_a^b dt\, L \tag{4.10.1}$$

is stationary, provided the endpoints are not varied:

$$\delta W_{ab} = \delta \int_a^b dt\, L = 0 \quad \text{if} \quad \delta t_a = \delta t_b = 0 \quad \text{and} \quad \delta x_a = \delta x_b = 0. \tag{4.10.2}$$

With $L = L(x, \dot{x}, t)$, where $\dot{x} = \dfrac{dx}{dt}$ is the velocity, this implies the familiar Euler–Lagrange equation

$$\frac{d}{dt}\frac{\partial L}{\partial \dot{x}} - \frac{\partial L}{\partial x} = 0. \tag{4.10.3}$$

For the standard form of the Lagrange function L,

$$L = \frac{M}{2}\dot{x}^2 - V(x) \qquad (M = \text{mass}), \tag{4.10.4}$$

it reads more explicitly

$$\frac{d}{dt}(M\dot{x}) + \frac{\partial V}{\partial x} = 0. \tag{4.10.5}$$

Since $-\dfrac{\partial V}{\partial x} = F$ is the force, this is of course just Newton's equation of motion in disguise.

Now, if we also allow endpoint variations, the varational principle says that all first-order changes come from the endpoints,

$$\delta W_{ab} = \delta \int_a^b dt\, L = G_b - G_a, \tag{4.10.6}$$

where G_a involves the infinitesimal endpoint variations δx_a and δt_a of the initial position and initial time, and G_b involves the variations δx_b and δt_b of the final position and the final time. To identify these *generators* G_a, G_b we introduce a path parameter τ, $0 \leq \tau \leq 1$, and regard the trajectory in x and t as a function of τ,

$$t = t(\tau), \quad x = x(\tau) \quad \text{with} \quad \begin{cases} t = t_a, \ x = x_a \text{ for } \tau = 0, \\ t = t_b, \ x = x_b \text{ for } \tau = 1. \end{cases} \tag{4.10.7}$$

Then

$$\dot{x} = \frac{\mathrm{d}x/\mathrm{d}\tau}{\mathrm{d}t/\mathrm{d}\tau} \equiv \frac{x'}{t'} \quad \text{and} \quad \mathrm{d}t = \mathrm{d}\tau \frac{\mathrm{d}t}{\mathrm{d}\tau} = \mathrm{d}\tau\, t', \tag{4.10.8}$$

and we write

$$W_{ab} = \int_0^1 \mathrm{d}\tau\, t' L(x, \dot{x}, t). \tag{4.10.9}$$

Now vary $t(\tau)$ and $x(\tau)$:

$$\begin{aligned} t(\tau) &\to t(\tau) + \delta t(\tau), \\ x(\tau) &\to x(\tau) + \delta x(\tau), \\ t'(\tau) &\to t'(\tau) + \delta t'(\tau) = t'(\tau) + \frac{\mathrm{d}}{\mathrm{d}\tau} \delta t(\tau), \end{aligned} \tag{4.10.10}$$

to get

$$\delta W_{ab} = \int_0^1 \mathrm{d}\tau \left(\delta t' L + t' \delta x \frac{\partial L}{\partial x} + t' \delta \dot{x} \frac{\partial L}{\partial \dot{x}} + t' \delta t \frac{\partial L}{\partial t} \right) \tag{4.10.11}$$

where

$$\delta \dot{x} = \delta \frac{x'}{t'} = \frac{\delta x'}{t'} - \frac{x' \delta t'}{t'^2} = \frac{1}{t'} \left(\delta x' - \delta t'\, \dot{x} \right). \tag{4.10.12}$$

With

$$\delta t' L = \frac{\mathrm{d}}{\mathrm{d}\tau} (\delta t L) - \delta t \frac{\mathrm{d}}{\mathrm{d}\tau} L \tag{4.10.13}$$

and

$$t'\delta\dot{x}\frac{\partial L}{\partial\dot{x}} = \delta x'\frac{\partial L}{\partial\dot{x}} - \delta t'\,\dot{x}\frac{\partial L}{\partial\dot{x}}$$
$$= \frac{d}{d\tau}\left(\delta x\frac{\partial L}{\partial\dot{x}} - \delta t\,\dot{x}\frac{\partial L}{\partial\dot{x}}\right)$$
$$- \delta x\frac{d}{d\tau}\frac{\partial L}{\partial\dot{x}} + \delta t\frac{d}{d\tau}\left(\dot{x}\frac{\partial L}{\partial\dot{x}}\right) \qquad (4.10.14)$$

we arrive at

$$\delta W_{ab} = \int_0^1 d\tau\left[\frac{d}{d\tau}\left(\delta x\frac{\partial L}{\partial\dot{x}} - \delta t\left(\dot{x}\frac{\partial L}{\partial\dot{x}} - L\right)\right)\right.$$
$$+ \delta t\left(-\frac{d}{d\tau}L + t'\frac{\partial L}{\partial t} + \frac{d}{d\tau}\left(\dot{x}\frac{\partial L}{\partial\dot{x}}\right)\right)$$
$$\left. + \delta x\left(t'\frac{\partial L}{\partial x} - \frac{d}{d\tau}\frac{\partial L}{\partial\dot{x}}\right)\right]. \qquad (4.10.15)$$

The total τ derivative gives the endpoint contributions

$$\delta W_{ab} = \left[\delta x\frac{\partial L}{\partial\dot{x}} - \delta t\left(\dot{x}\frac{\partial L}{\partial\dot{x}} - L\right)\right]\Bigg|_a^b = G_b - G_a \qquad (4.10.16)$$

with

$$G = \delta x\frac{\partial L}{\partial\dot{x}} - \delta t\left(\dot{x}\frac{\partial}{\partial\dot{x}} - L\right) = \delta x\,P - \delta t\,H\,,$$

which identifies the *momentum* $P = \dfrac{\partial L}{\partial\dot{x}}$ and the energy, or the *Hamilton function*,

$$H = \dot{x}\frac{\partial L}{\partial\dot{x}} - L\,. \qquad (4.10.17)$$

The internal contributions that are proportional to the independent infinitesimal path variations δx and δt must vanish, so that the δx term implies

$$t'\frac{\partial L}{\partial x} - \frac{d}{d\tau}\frac{\partial L}{\partial\dot{x}} = 0\,. \qquad (4.10.18)$$

With $\dfrac{d}{d\tau} = t'\dfrac{d}{dt}$ this is the Euler–Lagrange equation of motion (4.10.3), as it should be.

The generator

$$G = \delta x\,P - \delta t\,H \qquad (4.10.19)$$

provides $P = \dfrac{\partial L}{\partial x}$ and $H = \dot{x}\dfrac{\partial L}{\partial \dot{x}} - L$, which are just the usual kinetic momentum

$$P = M\dot{x} \qquad (4.10.20)$$

and the usual energy,

$$H = \frac{1}{2}M\dot{x}^2 + V(x)\,, \qquad (4.10.21)$$

for the Lagrange function (4.10.4).

We recognize in (4.10.19) the same formal structure as the one in (4.9.2), only that we are talking about the *momentum operator* and the *Hamilton operator* there, with the same fundamental minus sign between them. Note further that the response of H to arbitrary variations,

$$\delta H = \delta\dot{x}\,P + \dot{x}\delta P - \delta x\frac{\partial L}{\partial x} - \underbrace{\delta\dot{x}\,\frac{\partial L}{\partial\dot{x}}}_{=P} - \delta t\frac{\partial L}{\partial t}$$

$$= \dot{x}\,\delta P - \delta x\frac{\partial L}{\partial x} - \delta t\frac{\partial L}{\partial t} \qquad (4.10.22)$$

shows that the natural variables of H are P, x, and t,

$$H = H(P, x, t), \qquad (4.10.23)$$

whereas the natural variables of L are x, $\dot{x}$, and t. We must therefore express H as a a function of x, P, and t, which results in

$$H = \frac{1}{2M}P^2 + V(x) \qquad (4.10.24)$$

for the L in (4.10.4).

4.11 Hamilton operator, Schrödinger equation

We borrow this structure from classical mechanics and conjecture that the Hamilton operator of an atom of mass M in motion under the influence of potential energy V is given by

$$H = \frac{1}{2M}P^2 + V(X)\,. \qquad (4.11.1)$$

In particular, for force-free motion ("free particle") we only have the kinetic energy term,

$$H = \frac{1}{2M} P^2 \,. \tag{4.11.2}$$

The Schrödinger equation then reads

$$i\hbar \frac{\partial}{\partial t} \langle \ldots, t| = \langle \ldots, t| \frac{1}{2M} P^2 \tag{4.11.3}$$

for a general bra. Depending on the quantum numbers the ellipsis refers to, it becomes a numerical statement about time-dependent wave functions, with a large variety of appearances. For position wave functions, for example, we get

$$i\hbar \frac{\partial}{\partial t} \psi(x,t) = -\frac{\hbar^2}{2M} \frac{\partial^2}{\partial x^2} \psi(x,t) \quad \text{with} \quad \psi(x,t) = \langle x,t| \;\rangle \tag{4.11.4}$$

whereas

$$i\hbar \frac{\partial}{\partial t} \psi(p,t) = \frac{p^2}{2M} \psi(p,t) \quad \text{with} \quad \psi(p,t) = \langle p,t| \;\rangle \tag{4.11.5}$$

applies to momentum wave functions. They need to be solved subject to initial conditions that specify the wave functions at $t = 0$, say,

$$\langle x, t = 0| \;\rangle = \psi_0(x) \,,$$
$$\langle p, t = 0| \;\rangle = \psi_0(p) \,. \tag{4.11.6}$$

Since we can always switch from position wave functions to momentum wave functions and vice versa, solving one of the equations is tantamount to solving the other.

Before proceeding, let us recognize that these differential equations for the evolution of the wave functions for force-free motion are just particular cases of a much more general set of equations, namely the differential equations for arbitrary Hamilton operators

$$H(t) = H\big(P(t), X(t), t\big) \tag{4.11.7}$$

that have a dynamical time dependence because they are functions of the dynamical variables $P(t)$ and $X(t)$, and may have an additional parametric time dependence, if the functional dependence on P and X changes in time. Most typical is the form

$$H = \frac{1}{2M} P^2 + V(X,t) \,, \tag{4.11.8}$$

the sum of the standard kinetic energy $P^2/(2M)$ and situation-dependent potential energy $V(X,t)$ that might possess a parametric t dependence. Whatever the actual form of H as a function of P, X, and t, we get

$$i\hbar\frac{\partial}{\partial t}\psi(x,t) = i\hbar\frac{\partial}{\partial t}\langle x,t| \ \rangle = \langle x,t|H(P(t),X(t),t)| \ \rangle$$

$$\underset{\frac{\hbar}{i}\frac{\partial}{\partial x}}{\downarrow} \quad \underset{x}{\downarrow}$$

$$= H\Big(\frac{\hbar}{i}\frac{\partial}{\partial x},x,t\Big)\underbrace{\langle x,t| \ \rangle}_{=\psi(x,t)} \tag{4.11.9}$$

or

$$i\hbar\frac{\partial}{\partial t}\psi(x,t) = H\Big(\frac{\hbar}{i}\frac{\partial}{\partial x},x,t\Big)\psi(x,t) \tag{4.11.10}$$

for the position wave function. This differential equation, formally obtained by replacing $X \to x$, $P \to \frac{\hbar}{i}\frac{\partial}{\partial x}$ in $H(P,X,t)$ to turn the Hamilton operator into a differential operator, is frequently referred to as *the* Schrödinger equation. It is, in fact, the numerical version of the Schrödinger equation that is most frequently used in practice, but nonetheless it is only one of many Schrödinger equations. In particular, we have

$$i\hbar\frac{\partial}{\partial t}\psi(p,t) = i\hbar\frac{\partial}{\partial t}\langle p,t| \ \rangle = \langle p,t|H(P(t),X(t),t)| \ \rangle$$

$$\underset{p}{\downarrow} \quad \underset{i\hbar\frac{\partial}{\partial p}}{\downarrow}$$

$$= H\Big(p,i\hbar\frac{\partial}{\partial p},t\Big)\underbrace{\langle p,t| \ \rangle}_{=\psi(p,t)} \tag{4.11.11}$$

or

$$i\hbar\frac{\partial}{\partial t}\psi(p,t) = H\Big(p,i\hbar\frac{\partial}{\partial p},t\Big)\psi(p,t) \tag{4.11.12}$$

for the momentum wave function $\psi(p,t) = \langle p,t| \ \rangle$. And there are many more Schrödinger equations for other choices of wave functions.

The popularity of the Schrödinger equation for $\psi(x,t)$ originates in the circumstance that, usually, the momentum dependence of the Hamilton operator is quite simple, such as the one in (4.11.8), whereas the position dependence can be rather complicated. For Hamilton operators of this standard structure, we get

$$i\hbar\frac{\partial}{\partial t}\psi(x,t) = \left[-\frac{\hbar^2}{2M}\frac{\partial^2}{\partial x^2} + V(x,t)\right]\psi(x,t). \tag{4.11.13}$$

By all counts this must be the numerical version of the Schrödinger equation that is printed most often by far.

4.12 Time transformation function

Generally speaking, to solve such a Schrödinger equation means to express the wave function at the later time t in terms of the wave function at the initial time $t = 0$. That is, for example,

$$\psi(x,t) = \langle x,t| \ \rangle = \langle x,t| \int dx' \ |x',0\rangle \underbrace{\langle x',0| \ \rangle}_{=\,\psi_0(x')}$$

$$= \int dx' \ \langle x,t|x',0\rangle \, \psi_0(x') \tag{4.12.1}$$

or

$$\psi(x,t) = \langle x,t| \ \rangle = \langle x,t| \int dp \ |p,0\rangle \underbrace{\langle p,0| \ \rangle}_{=\,\psi_0(p)}$$

$$= \int dp \ \langle x,t|p,0\rangle \, \psi_0(p) \tag{4.12.2}$$

if we express the position wave function at t in terms of either the position wave function at $t = 0$ or the momentum wave function at $t = 0$. In view of

$$\psi_0(x) = \int dp \ \langle x,0|p,0\rangle \underbrace{\langle p,0| \ \rangle}_{=\,\psi_0(p)}$$

$$= \int dp \ \frac{e^{ixp/\hbar}}{\sqrt{2\pi\hbar}}\psi_0(p) \tag{4.12.3}$$

and its inverse relation

$$\psi_0(p) = \int dp \ \frac{e^{-ixp/\hbar}}{\sqrt{2\pi\hbar}}\psi_0(x), \tag{4.12.4}$$

we can go from one formulation to the other quite easily. So whether we calculate the *xx time-transformation function* $\langle x, t | x', 0 \rangle$ or the *xp time-transformation function* $\langle x, t | p, 0 \rangle$ or perhaps another one, does not matter that much, because we can turn one into the other, as illustrated by

$$\langle x, t | x', 0 \rangle = \int dp \, \langle x, t | p, 0 \rangle \langle p, 0 | x', 0 \rangle$$

$$= \int dp \, \langle x, t | p, 0 \rangle \frac{e^{-ipx'/\hbar}}{\sqrt{2\pi\hbar}} , \qquad (4.12.5)$$

which tells us how to obtain $\langle x, t | x', 0 \rangle$ when we know $\langle x, t | p, 0 \rangle$.

    Since

$$i\hbar \frac{\partial}{\partial t} \langle x, t | = \langle x, t | H\big(P(t), X(t), t\big)$$

$$= H\Big(\frac{\hbar}{i} \frac{\partial}{\partial x}, x, t\Big) \langle x, t | \qquad (4.12.6)$$

all transformation functions $\langle x, t | \ldots, 0 \rangle$ obey the same differential Schrödinger equation. What distinguishes them from each other are the corresponding initial conditions. For $\langle x, t | x', 0 \rangle$ we have a δ function,

$$\langle x, t | x', 0 \rangle \rightarrow \delta(x - x') \quad \text{as} \quad t \rightarrow 0, \qquad (4.12.7)$$

whereas we have a plane wave for $\langle x, t | p, 0 \rangle$,

$$\langle x, t | p, 0 \rangle \rightarrow \frac{e^{ixp/\hbar}}{\sqrt{2\pi\hbar}} \quad \text{as} \quad t \rightarrow 0. \qquad (4.12.8)$$

4-13   State the initial conditions for the *px* time-transformation function $\langle p, t | x, 0 \rangle$ and the *pp* time-transformation function $\langle p, t | p', 0 \rangle$.

It is often a matter of sheer convenience that makes us prefer one time transformation function over the other.

Chapter 5

Elementary Examples

5.1 Force-free motion

5.1.1 *Time-transformation functions*

Now, after these remarks of a quite general nature, let us return to the problem of finding the solution to the Schrödinger equation for an atom in force-free motion,

$$i\hbar\frac{\partial}{\partial t}\psi(x,t) = -\frac{\hbar^2}{2M}\frac{\partial^2}{\partial x^2}\psi(x,t)\,. \qquad (5.1.1)$$

It is convenient to first consider the xp transformation function, for which

$$i\hbar\frac{\partial}{\partial t}\langle x,t|p,0\rangle = \langle x,t|H|p,0\rangle \qquad (5.1.2)$$

with

$$H = \frac{1}{2M}P(t)^2 = \frac{1}{2M}P(0)^2 \qquad (5.1.3)$$

so that

$$H|p,0\rangle = |p,0\rangle\frac{p^2}{2M}\,. \qquad (5.1.4)$$

We recall here that a Hamilton operator that has no parametric time dependence, does not depend on time at all, $H(P(t),X(t)) = H(P(0),X(0))$. We shall return to this matter below, being content right now by the simplification thus achieved,

$$i\hbar\frac{\partial}{\partial t}\langle x,t|p,0\rangle = \langle x,t|p,0\rangle\frac{p^2}{2M}\,. \qquad (5.1.5)$$

This is an elementary differential equation, with the solution

$$\langle x, t | p, 0 \rangle = \frac{e^{ixp/\hbar}}{\sqrt{2\pi\hbar}} e^{-\frac{i}{\hbar}\frac{p^2}{2M}t} \tag{5.1.6}$$

where the time-independent prefactor incorporates the initial condition (4.12.8). Accordingly, we can now get the position wave function at time t from the momentum wave function at time $t = 0$,

$$\begin{aligned}
\psi(x, t) &= \int dp \, \langle x, t | p, 0 \rangle \langle p, 0 | \, \rangle \\
&= \int dp \, \frac{e^{ixp/\hbar}}{\sqrt{2\pi\hbar}} e^{-\frac{i}{\hbar}\frac{p^2}{2M}t} \psi_0(p) \,.
\end{aligned} \tag{5.1.7}$$

We use

$$\begin{aligned}
\psi_0(p) = \langle p, 0 | \, \rangle &= \int dx \, \langle p, 0 | x, 0 \rangle \langle x, 0 | \, \rangle \\
&= \int dx \, \frac{e^{-ipx/\hbar}}{\sqrt{2\pi\hbar}} \psi_0(x)
\end{aligned} \tag{5.1.8}$$

to introduce the position wave function at time $t = 0$,

$$\begin{aligned}
\psi(x, t) &= \int dp \, \frac{e^{ixp/\hbar}}{\sqrt{2\pi\hbar}} e^{-\frac{i}{\hbar}\frac{p^2}{2M}t} \int dx' \, \frac{e^{-ipx'/\hbar}}{\sqrt{2\pi\hbar}} \psi_0(x') \\
&= \int dx' \int dp \, \frac{e^{i(x-x')p/\hbar}}{2\pi\hbar} e^{-\frac{i}{\hbar}\frac{p^2}{2M}t} \psi_0(x')
\end{aligned} \tag{5.1.9}$$

which we compare with

$$\psi(x, t) = \int dx' \, \langle x, t | x', 0 \rangle \, \psi_0(x') \tag{5.1.10}$$

to read off that

$$\langle x, t | x', 0 \rangle = \int dp \, \frac{e^{i(x-x')p/\hbar}}{2\pi\hbar} e^{-\frac{i}{\hbar}\frac{p^2}{2M}t} \,. \tag{5.1.11}$$

Of course, this is nothing but the relation

$$\langle x, t | x', 0 \rangle = \int dp \, \langle x, t | p, 0 \rangle \langle p, 0 | x', 0 \rangle \tag{5.1.12}$$

with the particular force-free version of $\langle x, t | p, 0 \rangle$ inserted along with the universal $\langle p, 0 | x', 0 \rangle = e^{-ipx'/\hbar}/\sqrt{2\pi\hbar}$.

After completing the square in the exponent,

$$\frac{i}{\hbar}\frac{p^2}{2M}t + \frac{i}{\hbar}(x-x')p = -\frac{i}{\hbar}\frac{t}{2M}\left(p - \frac{M}{t}(x-x')\right)^2 + \frac{i}{\hbar}\frac{M}{2t}(x-x')^2,$$

(5.1.13)

the remaining integral is of gaussian type and thus easily evaluated,

$$\langle x,t|x',0\rangle = \int dp \frac{1}{2\pi\hbar} e^{-\frac{i}{\hbar}\frac{t}{2M}\left[p - \frac{M}{t}(x-x')\right]^2} e^{\frac{i}{\hbar}\frac{M}{2t}(x-x')^2}$$

$$= \frac{1}{2\pi\hbar}\sqrt{\frac{\pi}{\frac{i}{\hbar}\frac{t}{2M}}} e^{\frac{i}{\hbar}\frac{M}{2t}(x-x')^2},$$

(5.1.14)

or finally

$$\langle x,t|x',0\rangle = \sqrt{\frac{M}{i2\pi\hbar t}} e^{\frac{i}{\hbar}\frac{M}{2t}(x-x')^2}.$$

(5.1.15)

The initial condition (4.12.7) is correctly incorporated, inasmuch as we can verify the integral properly

$$\int dx' \langle x,t|x',0\rangle = \sqrt{\frac{M}{i2\pi\hbar t}}\sqrt{\frac{\pi}{-\frac{i}{\hbar}\frac{M}{2t}}} = 1,$$

(5.1.16)

and note that the function oscillates arbitrarily rapidly for $x' \neq x$ and very small t, so that any x' integration with a smooth function will not have noticeable contributions beyond the immediate vicinity of $x' \cong x$, where the function is very large for small t.

5-1 Find the time-transformation function $\langle p,t|p',0\rangle$ from momentum states to momentum states.

5.1.2 *"Spreading" of the wave function*

The constancy in time of the Hamilton operator $H = \frac{1}{2M}P^2$ implies a constancy in time of the momentum operator P, but that must also be evident from Heisenberg's equation of motion for $P(t)$. Indeed,

$$\frac{d}{dt}P(t) = \frac{1}{i\hbar}[P(t),H] = \frac{1}{i\hbar}\left[P(t), \frac{1}{2M}P(t)^2\right] = 0.$$

(5.1.17)

And so

$$P(t) = P(0).$$

(5.1.18)

Likewise we have

$$\frac{d}{dt}X(t) = \frac{1}{i\hbar}\left[X(t), \frac{1}{2M}P(t)^2\right] = \frac{1}{M}P(t),$$ (5.1.19)

where the right-hand side — the velocity operator — is constant,

$$\frac{d}{dt}X(t) = \frac{1}{M}P(0),$$ (5.1.20)

so that we can integrate immediately to arrive at

$$X(t) = X(0) + \frac{t}{M}P(0).$$ (5.1.21)

Note that these equations look exactly the same as their classical counterparts, but now they are equations about operators, rather than numbers, depending on time.

Upon taking expectation values, we obtain numerical statements,

$$\langle P(t)\rangle = \langle P(0)\rangle,$$
$$\langle X(t)\rangle = \langle X(0)\rangle + \frac{t}{M}\langle P(0)\rangle,$$ (5.1.22)

which are even closer counterparts to the classical analogs. Aiming at an evaluation of the position and momentum spreads, $\delta X(t)$ and $\delta P(t)$, we also find the expectation values of the squared operators,

$$\langle P(t)^2\rangle = \langle P(0)^2\rangle,$$
$$\langle X(t)^2\rangle = \left\langle \left(X(0) + \frac{t}{M}P(0)\right)^2\right\rangle$$
$$= \langle X(0)^2\rangle + \frac{t^2}{M^2}\langle P(0)^2\rangle$$
$$+ \frac{t}{M}\left\langle \left(X(0)P(0) + P(0)X(0)\right)\right\rangle.$$ (5.1.23)

Accordingly, for P we get

$$\delta P(t)^2 = \langle P(t)^2\rangle - \langle P(t)\rangle^2$$
$$= \langle P(0)^2\rangle - \langle P(0)\rangle^2 = \delta P(0)^2,$$ (5.1.24)

as we anticipated, of course: no force, no change in momentum whatsoever, and for X we obtain

$$\delta X(t)^2 = \left\langle X(t)^2 \right\rangle - \left\langle X(t) \right\rangle^2$$

$$= \left\langle X(0)^2 \right\rangle - \left\langle X(0) \right\rangle^2 + \frac{t^2}{M^2} \left(\left\langle P(0)^2 \right\rangle - \left\langle P(0) \right\rangle^2 \right)$$

$$+ \frac{t}{M} \left(\left\langle \left(X(0)P(0) + P(0)X(0) \right) \right\rangle - 2 \left\langle X(0) \right\rangle \left\langle P(0) \right\rangle \right)$$

(5.1.25)

or

$$\delta X(t)^2 = \delta X(0)^2 + \left(\frac{t}{M} \delta P(0) \right)^2$$

$$+ \frac{2t}{M} \left(\left\langle \frac{X(0)P(0) + P(0)X(0)}{2} \right\rangle - \left\langle X(0) \right\rangle \left\langle P(0) \right\rangle \right).$$

(5.1.26)

This tells us that the spread in position *does* depend on time, and more: it always grows linearly with t for very late times t, because for large t, the dominating term is the one $\propto t^2$, so that

$$\delta X(t) \cong \frac{t}{M} \delta P(0) \qquad \text{for very late times } t. \qquad (5.1.27)$$

This phenomenon is colloquially known as the "spreading of the wave function". It is, in fact, something rather natural, and something rather obvious: Because we do not know the initial velocity precisely, but only with an error measured by $\delta P/M$, our predictions about the object's position will become less precise in time. This is *all* there is to it. Do not imagine that the atom becomes spread out, what is spreading is its wave function, not the atom itself. When you look for it, you will always find the atom in one complete piece at a certain place, but if you wait too long before looking, you cannot guess very well where you will find the atom. The imprecision is in our capability of predicting the outcome of a position measurement.

5-2 Take the minimum-uncertainty wave function of (4.8.10) as the initial wave function $\psi_0(x)$ and calculate $\left\langle \left(X(0)P(0) + P(0)X(0) \right) \right\rangle$. Then, with the previously established values for $\left\langle X(0) \right\rangle$, $\left\langle X(0)^2 \right\rangle$, as well as $\left\langle P(0) \right\rangle, \left\langle P(0)^2 \right\rangle$, determine $\delta X(t)$. How long does it take until $\delta X(t)^2$ is twice $\delta X(0)^2$? Express this "doubling time" in terms of $\delta X(0)$ and $\delta P(0)$ without involving $\hbar$.

5-3　Express $\left\langle \frac{1}{2}\big(X(t)P(t) + P(t)X(t)\big)\right\rangle - \langle X(t)\rangle\,\langle P(t)\rangle$ in terms of expectation values at $t = 0$. What do you get, in particular, for the initial wave function of the preceding exercise?

We can see this "spreading of the wave function" also — of course — by looking at the wave function itself rather than the expectation values that go with it. Fitting to Exercises 5-2 and 5-3, we take

$$\psi_0(x) = \frac{(2\pi)^{-\frac{1}{4}}}{\sqrt{\delta X}}\, \mathrm{e}^{-(\frac{1}{2}x/\delta X)^2} \tag{5.1.28}$$

and have

$$\psi(x,t) = \int \mathrm{d}x' \sqrt{\frac{M}{\mathrm{i}2\pi\hbar t}}\, \mathrm{e}^{\frac{\mathrm{i}}{\hbar}\frac{M}{2t}(x - x')^2} \frac{(2\pi)^{-\frac{1}{4}}}{\sqrt{\delta X}}\, \mathrm{e}^{-(\frac{1}{2}x'/\delta X)^2} \,. \tag{5.1.29}$$

The exponent is a quadratic function of x', so this is another gaussian integral, which we can evaluate explicitly without much ado. First, we complete a square, for which the identity

$$a(x' - x_1)^2 + b(x' - x_2)^2 = (a + b)\left(x' - \frac{ax_1 + bx_2}{a + b}\right)^2 - \frac{ab}{a + b}(x_1 - x_2)^2 \tag{5.1.30}$$

is useful. Here we use it for

$$a = \left(\frac{1}{2\delta X}\right)^2, \qquad b = -\frac{\mathrm{i}}{\hbar}\frac{M}{2t}, \qquad x_1 = 0, \qquad x_2 = x, \tag{5.1.31}$$

and get

$$\left(\frac{x}{2\delta X}\right)^2 - \frac{\mathrm{i}}{\hbar}\frac{M}{2t}(x - x')^2 = \left(\frac{1}{(2\delta X)^2} - \frac{\mathrm{i}}{\hbar}\frac{M}{2t}\right)\big[x' - (\text{something})\big]^2$$
$$- \frac{x^2}{(2\delta X)^2 + \mathrm{i}2\hbar t/M}\,. \tag{5.1.32}$$

We do not have to work out the "something", since it just amounts to a shift of the integration variable and has no relevance for the value of the x' integral, and we used

$$\frac{ab}{a + b} = \frac{1}{1/a + 1/b} \qquad \text{(harmonic mean)} \tag{5.1.33}$$

in the last term. After writing

$$(2\delta X)^2 + \underbrace{\mathrm{i}2\hbar\, t/M}_{=4\delta X\,\delta P} = (2\delta X)^2 \underbrace{\left(1 + \mathrm{i}\frac{t}{M}\frac{\delta P}{\delta X}\right)}_{\equiv 1/\epsilon(t)} \qquad (5.1.34)$$

or

$$\frac{1}{a} + \frac{1}{b} = \frac{1}{a\epsilon}, \qquad (5.1.35)$$

we now note that the x independent factor resulting from the gaussian integral is

$$\underbrace{\sqrt{\frac{M}{\mathrm{i}2\pi\hbar t}}}_{=\sqrt{b/\pi}}\frac{(2\pi)^{-\frac{1}{4}}}{\sqrt{\delta X}}\sqrt{\frac{\pi}{a+b}} = \frac{(2\pi)^{-\frac{1}{4}}}{\delta X}\sqrt{\frac{b}{a+b}}$$

$$= \frac{(2\pi)^{-\frac{1}{4}}}{\delta X}\sqrt{\frac{1}{a}\left(\frac{1}{a}+\frac{1}{b}\right)^{-1}}$$

$$= \frac{(2\pi)^{-\frac{1}{4}}}{\delta X}\sqrt{\epsilon}. \qquad (5.1.36)$$

Accordingly,

$$\psi(x,t) = \frac{(2\pi)^{-\frac{1}{4}}}{\sqrt{\delta X/\epsilon(t)}}\, \mathrm{e}^{-\epsilon(t)\left(\frac{1}{2}x/\delta X\right)^2}$$

$$\text{with}\quad \epsilon(t) = \left(1 + \mathrm{i}\frac{t}{M}\frac{\delta P}{\delta X}\right)^{-1}. \qquad (5.1.37)$$

The wave function keeps its gaussian shape in the course of time, but acquires an imaginary part. The "spreading" becomes particularly visible in the associated probability density $|\psi(x,t)|^2$,

$$|\psi(x,t)|^2 = \frac{1}{\sqrt{2\pi}}\frac{1}{\delta X}|\epsilon(t)|\, \mathrm{e}^{-\frac{1}{2}(x/\delta X)^2\,\mathrm{Re}(\epsilon(t))}, \qquad (5.1.38)$$

where we meet $|\epsilon(t)|$ and $\mathrm{Re}(\epsilon(t))$, given by

$$|\epsilon(t)|^2 = \mathrm{Re}(\epsilon(t)) = \left[1 + \left(\frac{t\delta P}{M\delta X}\right)^2\right]^{-1}. \qquad (5.1.39)$$

Therefore,

$$|\psi(x,t)|^2 = \frac{1}{\sqrt{2\pi}}\frac{1}{\delta X(t)}\, \mathrm{e}^{-\frac{1}{2}(x/\delta X(t))^2} \qquad (5.1.40)$$

with

$$\delta X(t) = \frac{\delta X}{|\epsilon(t)|} = \frac{\delta X}{\sqrt{\mathrm{Re}(\epsilon(t))}}$$

$$= \delta X \sqrt{1 + \left(\frac{t\delta P}{M\delta X}\right)^2} = \sqrt{(\delta X)^2 + \left(\frac{t}{M}\delta P\right)^2}. \quad (5.1.41)$$

Not surprisingly, this agrees with the spread $\delta X(t)$ found earlier (see Exercise 5-2 on page 145). Upon comparing with the general expression (5.1.26), we conclude that for the chosen $\psi_0(x)$ the term $\propto t$ vanishes, that is

$$\left\langle \frac{X(0)P(0) + P(0)X(0)}{2} \right\rangle = \left\langle X(0) \right\rangle \left\langle P(0) \right\rangle, \quad (5.1.42)$$

another observation made already in Exercise 5-3.

The "spreading" has yet another aspect in addition to the growing spread in position. Let us ask: Since $\left\langle X(t) \right\rangle = 0$ and $\left\langle P(t) \right\rangle = 0$ for all t, the atom appears to be stationary, so what is the probability that the dynamical evolution does not change anything? In other, more technical words: What is the probability of finding the atom, after the lapse of time t, in a state that has, at that later time, the wave function that the atom had initially? That is,

$$\mathrm{prob(no\ change)} = \left| \int \mathrm{d}x \; \underbrace{\psi_0(x)^*}\; \underbrace{\psi(x,t)} \right|^2 = ?. \quad (5.1.43)$$

$$\text{before the "spreading"} \qquad \text{after the "spreading"}$$

Since we are dealing with a product of gaussian wave functions, we can evaluate the integral quite easily,

$$\int \mathrm{d}x \; \psi_0(x)^*\psi(x,t) = \int \mathrm{d}x \; \frac{(2\pi)^{-\frac{1}{4}}}{\sqrt{\delta X}} e^{-(\frac{1}{2}x/\delta X)^2} \frac{(2\pi)^{-\frac{1}{4}}}{\sqrt{\delta X/\epsilon}} e^{-\epsilon(\frac{1}{2}x/\delta X)^2}$$

$$= \frac{1}{\sqrt{2\pi/\epsilon}\,\delta X} \sqrt{\frac{\pi}{\frac{1+\epsilon}{(2\delta X)^2}}} = \sqrt{\frac{2\epsilon}{1+\epsilon}}, \quad (5.1.44)$$

and arrive at

$$\mathrm{prob(no\ change)} = \left| \frac{2\epsilon}{1+\epsilon} \right| = \sqrt{\frac{4|\epsilon|^2}{1 + 2\,\mathrm{Re}(\epsilon) + |\epsilon|^2}}$$

$$= \sqrt{\frac{4|\epsilon|^2}{1 + 3|\epsilon|^2}} = \left(\frac{3}{4} + \frac{1}{4|\epsilon|^2} \right)^{-\frac{1}{2}} \quad (5.1.45)$$

where $\text{Re}(\epsilon) = |\epsilon|^2$ is recalled. We insert the expression for $|\epsilon|^2$ from above and get

$$\text{prob(no change)} = \left[1 + \left(\frac{t\,\delta P}{2M\delta X} \right)^2 \right]^{-\frac{1}{2}}. \tag{5.1.46}$$

5.1.3 Long-time and short-time behavior

Two details are remarkable, namely, what happens for very short times and what for very long times, meaning

$$\text{very short times:} \quad \left(\frac{t\,\delta P}{2M\delta X} \right) \ll 1\,,$$

$$\text{very long times:} \quad \left(\frac{t\,\delta P}{2M\delta X} \right) \gg 1\,. \tag{5.1.47}$$

For very long times, we note that "decay" of this kind is not exponential, rather the probability of having not yet decayed,

$$\text{prob(no change after long time t)} \cong \frac{2M\delta X}{t\,\delta P} \propto \frac{1}{t}\,, \tag{5.1.48}$$

decreases with a power law, here t^{-1}. This is a generic feature in the sense that such long-time behavior cannot be exponential, it must always be slower, but the precise dependence on t varies from circumstance to circumstance. Roughly speaking one gets t^{-N} where N is the number of degrees of freedom that are involved. All of this is a consequence of the fact that all physical Hamilton operators are bounded from below — they have a state of lowest energy, a ground state.

The short-time behavior,

$$\text{prob(no change after a short time t)} \cong 1 - \frac{1}{2} \left(\frac{t\,\delta P}{2M\delta X} \right)^2$$

$$= 1 - \underbrace{(\cdots)}_{>0}\, t^2 \tag{5.1.49}$$

is also generic and very much so, because we always have a deviation that is proportional to t^2 and the positive coefficient that sets the time scale has a particular physical significance that we shall explore now.

Let us recall what we calculated. The probability (5.1.43) in question results from the probability amplitude

$$\int \mathrm{d}x \, \psi_0(x)^* \psi(x,t) = \int \mathrm{d}x \, \langle \, |x,0\rangle\langle x,t| \, \rangle$$

$$= \langle \, |\left(\int \mathrm{d}x \, |x,0\rangle\langle x,t|\right)| \, \rangle \equiv \langle \, |U_{0,t}| \, \rangle , \quad (5.1.50)$$

which is the expectation value of the unitary evolution operator

$$U_{0,t} = \int \mathrm{d}x \, |x,0\rangle\langle x,t| \qquad (5.1.51)$$

that links bras at one time to those at another time,

$$\langle x,0|U_{0,t} = \langle x,t| . \qquad (5.1.52)$$

5-4 Show that the *same* $U_{0,t}$ appears in

$$\langle p,0|U_{0,t} = \langle p,t|$$

and then conclude that $\langle \ldots, t| = \langle \ldots, 0|U_{0,t}$ for any corresponding quantum numbers.

The Schrödinger equation obeyed by $\langle x,t|$,

$$\mathrm{i}\hbar \frac{\partial}{\partial t} \langle x,t| = \langle x,t|H(t) , \qquad (5.1.53)$$

implies immediately that

$$U_{0,t} = \mathrm{e}^{-\mathrm{i}Ht/\hbar} \qquad (5.1.54)$$

provided that H does not depend on time,

$$\frac{\mathrm{d}}{\mathrm{d}t} H = \frac{\partial H}{\partial t} = 0 . \qquad (5.1.55)$$

The complications that arise for Hamilton operators with a parametric time dependence are of a technical nature, and we do not want to be distracted by them.

Accordingly, we have

$$\mathrm{prob(no\ change)} = \left| \left\langle \mathrm{e}^{-\mathrm{i}Ht/\hbar} \right\rangle \right|^2 \qquad (5.1.56)$$

which we now evaluate to second order in t. First note that

$$\left\langle e^{-iHt/\hbar} \right\rangle = \left\langle \left(1 - \frac{i}{\hbar} H t + \frac{1}{2} \left(-\frac{i}{\hbar} H t \right)^2 + \cdots \right) \right\rangle$$

$$= 1 - \frac{i}{\hbar} \langle H \rangle t - \frac{1}{2} \frac{1}{\hbar^2} \left\langle H^2 \right\rangle t^2 + \cdots, \qquad (5.1.57)$$

and then that

$$\left| \left\langle e^{-iHt/\hbar} \right\rangle \right|^2 = \left(1 - \frac{1}{2} \frac{1}{\hbar^2} \left\langle H^2 \right\rangle t^2 \right)^2 + \left(\frac{1}{\hbar} \langle H \rangle t \right)^2 + \cdots$$

$$= 1 - \frac{1}{\hbar^2} \left(\left\langle H^2 \right\rangle - \langle H \rangle^2 \right) t^2 + \cdots. \qquad (5.1.58)$$

We meet here the *energy spread*

$$\delta E = \delta H = \sqrt{\left\langle H^2 \right\rangle - \langle H \rangle^2} \qquad (5.1.59)$$

and so arrive at

$$\text{prob(no change for small } t) = 1 - \left(\frac{\delta E}{\hbar} t \right)^2 + O(t^4). \qquad (5.1.60)$$

This says that the time scale for the initial "decay" is set by the energy spread δE,

$$\text{prob(no change)} = 1 - \left(\frac{t}{T} \right)^2 + O(t^4) \quad \text{with} \quad T = \frac{\hbar}{\delta E}. \qquad (5.1.61)$$

A large energy spread leads to rapid decay, a small one to slow decay.

No decay at all,

$$\text{prob(no change after } t) = 1, \qquad (5.1.62)$$

requires $\delta E = 0$. Then the t^2 term disappears, but also all higher-order terms. For, $\delta E = 0$ implies that $| \rangle$ is an eigenstate of the Hamilton operator,

$$H | \rangle = | \rangle E, \qquad (5.1.63)$$

so that the outcome of an energy measurement can be predicted to be this eigenvalue E with certainty. Then

$$\left\langle e^{-iHt/\hbar} \right\rangle = e^{-iEt/\hbar} \qquad (5.1.64)$$

and prob(no change)$= \left| e^{-iEt/\hbar} \right|^2 = 1$, indeed there is no change at all.

Although it is quite obvious that, for any hermitian operator A, a vanishing spread $\delta A = 0$, implies that the ket in question is an eigenket of A, let us give another argument for this fact, an argument of a somewhat geometrical nature.

Given $A = A^\dagger$ and ket $|\ \rangle$ of unit length, $\langle\ |\ \rangle = 1$, then $A|\ \rangle$ is a new ket that, quite generally, has a component parallel to $|\ \rangle$ and one orthogonal to $|\ \rangle$,

$$A|\ \rangle = |\ \rangle a + |\perp\rangle b \tag{5.1.65}$$

with coefficients a, b and $\langle\ |\perp\rangle = 0, \langle\perp|\perp\rangle = 1$. The significance of a is immediate,

$$a = \langle\ |\left(A|\ \rangle\right) = \langle A\rangle\,, \tag{5.1.66}$$

it is the expectation value of A for ket $|\ \rangle$. It follows that a is real whereas b can be complex.

The significance of b is then revealed by considering the square of $A|\ \rangle$,

$$\left(\langle\ |A\rangle\left(A|\ \rangle\right) = \langle A^2\rangle = a^2 + b^*b\,, \tag{5.1.67}$$

so that

$$|b|^2 = \langle A^2\rangle - a^2 = \langle A^2\rangle - \langle A\rangle^2 = (\delta A)^2\,.$$

We thus have this picture

and it follows immediately that $\delta A = 0$ implies

$$A|\ \rangle = |\ \rangle a\,, \tag{5.1.68}$$

that is: $|\ \rangle$ is eigenket of A, and then $\langle e^{i\beta A}\rangle = e^{i\beta a}$ is a pure phase factor, $\left|\langle e^{i\beta A}\rangle\right| = 1$, for all real β. In the context of the above considerations of prob(no decay), it follows that there is no evolution whatsoever unless $\delta E \neq 0$. If the system is in an eigenstate of the Hamilton operator, $\delta E = 0$, then it does not evolve in time at all.

Comparing now the general result

$$1 - \left(\frac{\delta E}{\hbar} t\right)^2 + O(t^4) \tag{5.1.69}$$

for prob(no change) in (5.1.60) with the particular example of (5.1.49),

$$1 - \frac{1}{2}\left(\frac{t\,\delta P}{2M\delta X}\right)^2 + O(t^4) \quad \text{with} \quad \delta X\delta P = \frac{1}{2}\hbar, \tag{5.1.70}$$

we infer that here

$$\frac{1}{\hbar}\delta E = \frac{1}{\sqrt{8}}\frac{\delta P}{M\delta X} \quad \text{or} \quad \delta E = \frac{1}{\sqrt{2}}\frac{(\delta P)^2}{M}. \tag{5.1.71}$$

As a check of consistency, we verify that this is indeed the energy spread for $H = P^2/(2M)$ and

$$\psi_0(x) = \frac{(2\pi)^{-\frac{1}{4}}}{\sqrt{\delta X}}\, e^{-(\frac{1}{2}x/\delta X)^2}. \tag{5.1.72}$$

Since

$$\delta E = \sqrt{\left\langle H^2\right\rangle - \left\langle H\right\rangle^2} = \frac{1}{2M}\sqrt{\left\langle P^4\right\rangle - \left\langle P^2\right\rangle^2} \tag{5.1.73}$$

involves expectation values of powers of the momentum operator, it will be convenient to employ the momentum wave function

$$\psi_0(p) = \frac{(2\pi)^{-\frac{1}{4}}}{\sqrt{\delta P}}\, e^{-(\frac{1}{2}p/\delta P)^2} \tag{5.1.74}$$

rather than the position wave function (5.1.28). We have, in general,

$$\left\langle (P^2)^n\right\rangle = \int dp\,(p^2)^n \frac{1}{2\pi}\frac{1}{\delta P}\, e^{-\frac{1}{2}(p/\delta P)^2}, \tag{5.1.75}$$

which are gaussian-type integrals of the form

$$\int dp\,(p^2)^n\, e^{-\gamma p^2} = \left(-\frac{\partial}{\partial\gamma}\right)^n \int dp\, e^{-\gamma p^2} = \left(-\frac{\partial}{\partial\gamma}\right)^n \sqrt{\frac{\pi}{\gamma}}$$

$$= \frac{1}{2}\frac{3}{2}\cdots\frac{2n-1}{2}\sqrt{\frac{\pi}{\gamma^{2n+1}}} = \frac{(2n)!}{4^n\, n!}\sqrt{\frac{\pi}{\gamma^{2n+1}}}. \tag{5.1.76}$$

We need the $n = 1$ case in

$$\left\langle P^2\right\rangle = \frac{1}{\sqrt{2\pi}}\frac{1}{\delta P}\frac{1}{2}\sqrt{\frac{\pi}{\left(\frac{1}{2}/(\delta P)^2\right)^3}} = (\delta P)^2 \tag{5.1.77}$$

(which could have been anticipated), and the $n = 2$ case in

$$\left\langle P^4 \right\rangle = \frac{1}{\sqrt{2\pi}} \frac{1}{\delta P} \frac{3}{4} \sqrt{\pi \left(2(\delta P)^2\right)^5} = 3(\delta P)^4 . \qquad (5.1.78)$$

Together,

$$\delta(P^2) = \sqrt{\left\langle P^4 \right\rangle - \left\langle P^2 \right\rangle^2} = \sqrt{3(\delta P)^4 - \left((\delta P)^2\right)^2} = \sqrt{2}(\delta P)^2 , \quad (5.1.79)$$

and

$$\delta E = \frac{1}{2M} \sqrt{2}(\delta P)^2 = \frac{1}{\sqrt{2}} \frac{(\delta P)^2}{M} \qquad (5.1.80)$$

follows, indeed.

A final remark on these matters is the observation that we have, in a sense, seen the quadratic initial time dependence before, namely when discussing many intermediate Stern–Gerlach measurements in Section 2.6. The general statement that we can formulate now is about many intermediate control measurements asking "has some evolution happened?". If the total duration T is broken up into n intervals of duration $t = T/n$, then the probability of no change all together will be

$$\text{prob(no change in } T \text{, with } n \text{ controls)}$$
$$= \left[\text{prob(no change in } t = T/n)\right]^n . \qquad (5.1.81)$$

Now if n is so large that the quadratic short-time approximation applies to $t = T/n$, then

$$\text{prob(no change in } T \text{, with } n \text{ controls)} \cong \left[1 - \left(\frac{\delta E}{\hbar} \frac{T}{n}\right)^2\right]^n$$

$$\cong \left[e^{-\left(\frac{\delta E}{\hbar} \frac{T}{n}\right)^2}\right]^n = e^{-\frac{1}{n}\left(\delta E \frac{T}{\hbar}\right)^2} \cong 1 - \frac{1}{n}\left(\delta E \frac{T}{\hbar}\right)^2 . \quad (5.1.82)$$

This states an effective slowdown because the time constant for decay, $\sqrt{n}\hbar/\delta E$, is $\sqrt{n}$ times the original one. For n so large that $\frac{1}{n}\left(\delta E \frac{T}{\hbar}\right)^2$ is of the desired small size, the evolution is thus effectively halted by the many interrupting control measurements. As mentioned in Section 2.6, this "quantum Zeno effect" is a real physical phenomenon that has been demonstrated repeatedly in a variety of experiments.

5.1.4   *Interlude: General position-dependent force*

In the presence of a force, $F = -\frac{\partial}{\partial x}V(x)$, derived from a potential energy $V(x)$, the Hamilton operator is of the typical form

$$H = \frac{1}{2M}P^2 + V(X).$$ (5.1.83)

The Heisenberg equations of motion are then

$$\frac{\mathrm{d}}{\mathrm{d}t}X = \frac{1}{i\hbar}[X,H] = \frac{1}{i\hbar}\left[X, \frac{1}{2M}P^2\right] = \frac{1}{M}P$$ (5.1.84)

and

$$\frac{\mathrm{d}}{\mathrm{d}t}P = \frac{1}{i\hbar}[P,H] = \frac{1}{i\hbar}[P,V(X)] = -\frac{\partial V}{\partial X}(X)$$ (5.1.85)

or

$$\frac{\mathrm{d}}{\mathrm{d}t}P = F(X).$$ (5.1.86)

The latter commutator is an example of

$$[f(X),P] = i\hbar\frac{\partial f}{\partial X}(X),$$ (5.1.87)

that is: we differentiate with respect to X by taking the commutator with P. This equation can be proven in various ways, simplest perhaps by applying the operators to x bras,

$$\begin{aligned}
\langle x|[f(X),P] &= \langle x|(f(X)P - Pf(X)) \\
&= \left(f(x)\frac{\hbar}{i}\frac{\partial}{\partial x} - \frac{\hbar}{i}\underbrace{\frac{\partial}{\partial x}f(x)}_{=f(x)\frac{\partial}{\partial x} + \frac{\partial f}{\partial x}}\right)\langle x| \\
&= i\hbar\frac{\partial f(x)}{\partial x}\langle x| = \langle x|i\hbar\frac{\partial f(X)}{\partial X},
\end{aligned}$$ (5.1.88)

and then the commutator identity follows from the completeness of the bras $\langle x|$.

5-5 Show that $[X,g(P)] = i\hbar\frac{\partial g}{\partial P}(P)$, that is: we differentiate with respect to P by taking the commutator with X.

The pair of Heisenberg equations of motion (5.1.84) and (5.1.85) constitute, as a rule, a coupled set of *nonlinear* differential equation for the *operators* $X(t)$ and $P(t)$, because an arbitrary function $V(X)$ is involved, and we do not have methods for solving them explicitly. By "solving" we mean to express $X(t)$ and $P(t)$ in terms of $X(0)$ and $P(0)$, as we managed to do for the force-free particle case of $V \equiv 0$. There are, however, various methods for obtaining approximate solutions in a systematic manner, part and parcel of *perturbation theory*, which is beyond the scope of *Basic Matters* but is discussed in *Simple Systems* and *Perturbed Evolution*.

One can always retreat to the formal solutions

$$\begin{aligned}
X(t) &= \mathrm{e}^{\mathrm{i}Ht/\hbar} X(0)\, \mathrm{e}^{-\mathrm{i}Ht/\hbar}, \\
P(t) &= \mathrm{e}^{\mathrm{i}Ht/\hbar} P(0)\, \mathrm{e}^{-\mathrm{i}Ht/\hbar},
\end{aligned} \tag{5.1.89}$$

which solve the Heisenberg equations of motion, provided that H itself does not depend on time

$$\frac{\mathrm{d}H}{\mathrm{d}t} = 0, \tag{5.1.90}$$

which, as we know, is the requirement that there is no parametric time dependence in H, $\partial H/\partial t = 0$. More generally, any operator function $F\big(X(t), P(t)\big)$ can be related to its $t = 0$ version by

$$F\big(X(t), P(t)\big) = \mathrm{e}^{\mathrm{i}Ht/\hbar} F\big(X(0), P(0)\big)\, \mathrm{e}^{-\mathrm{i}Ht/\hbar}, \tag{5.1.91}$$

and the unitary evolution operator $\mathrm{e}^{-\mathrm{i}Ht/\hbar}$ also provides the time dependent bras,

$$\langle \ldots, t| = \langle \ldots, 0|\, \mathrm{e}^{-\mathrm{i}Ht/\hbar}, \tag{5.1.92}$$

as we have already observed in (5.1.52)–(5.1.54).

All of this directs our attention to (3.8.4), that is

$$\mathrm{e}^{-\mathrm{i}Ht/\hbar} = \sum_j |E_j\rangle\, \mathrm{e}^{-\mathrm{i}E_j t/\hbar} \langle E_j|, \tag{5.1.93}$$

where the sum is over the eigenvalues of H,

$$H|E_j\rangle = |E_j\rangle E_j, \qquad \langle E_j|H = E_j\langle E_j|, \tag{5.1.94}$$

physically speaking the *eigenenergies* of the system. This overwhelming importance for time evolution gives a very special role to the eigenvalues and eigenstates of the Hamilton operator and, therefore, a variety of techniques have been developed for determining the eigenenergies and the energy eigenstates with high precision. Some of these methods are discussed in *Simple Systems*. They are, of course, also applicable to other hermitian operators, but most of the time it is the Hamilton operator that is both of special interest and particularly difficult to deal with.

5.1.5 Energy eigenstates

For the force-free motion we have $H = P^2/(2M)$ and the time-independent Schrödinger equation (that is: the eigenvalue equation for H) reads

$$\frac{1}{2M}P^2|E\rangle = |E\rangle E. \tag{5.1.95}$$

Clearly, the eigenkets of the momentum operator P are also eigenkets of its square $P^2 = 2MH$ and, therefore, we may identify one with the other,

$$|E\rangle \propto |p\rangle \quad \text{with} \quad E = \frac{1}{2M}p^2. \tag{5.1.96}$$

We encounter here, in a simple context, a quite ubiquitous detail, namely that there is more than one state for a given energy. It is rather common that energy eigenvalues are *degenerate* by which we mean just that: there is more than one state for the given energy eigenvalue. Such degeneracies are always the consequence of some symmetry property. In the case of the force-free motion it is the *reflection symmetry*: it does not matter whether the atom moves to the left ($p < 0$) or to the right ($p > 0$).

We thus have two eigenstates for each energy eigenvalue $E > 0$,

$$|E, +\rangle \propto |p = +\sqrt{2ME}\rangle, \tag{5.1.97}$$

and

$$|E, -\rangle \propto |p = -\sqrt{2ME}\rangle. \tag{5.1.98}$$

All "+" kets are automatically orthogonal to the "−" kets,

$$\langle E, +|E', -\rangle = 0, \tag{5.1.99}$$

and since E is a continuous label (all $E > 0$ are permissible), we have to normalize the energy eigenkets to a δ function

$$\begin{aligned} \langle E, +|E', +\rangle &= \delta(E - E'), \\ \langle E, -|E', -\rangle &= \delta(E - E'). \end{aligned} \tag{5.1.100}$$

With this normalization, the identity is decomposed as

$$1 = \sum_{\alpha=\pm} \int_0^\infty dE\, |E, \alpha\rangle\langle E, \alpha|. \tag{5.1.101}$$

We verify this completeness relation by applying it to $\left|E',\beta\right\rangle$ with $E' > 0$ and $\beta = \pm$,

$$\int_0^\infty dE \sum_\alpha \underbrace{\left|E,\alpha\right\rangle\left\langle E,\alpha\middle|E',\beta\right\rangle}_{=\delta_{\alpha\beta}\delta(E-E')} = \left|E',\beta\right\rangle. \tag{5.1.102}$$

Writing,

$$\left|E,\pm\right\rangle = \left|p=\pm\sqrt{2ME}\right\rangle a_\pm(E) \tag{5.1.103}$$

with the normalizing factor $a_\pm(E)$ to be determined, we have

$$\begin{aligned}
\left\langle E,\alpha\middle|E',\beta\right\rangle &= a_\alpha(E)^*\left\langle p=\alpha\sqrt{2ME}\middle|p=\beta\sqrt{2ME'}\right\rangle a_\beta(E') \\
&= a_\alpha(E)^*\underbrace{\delta\left(\alpha\sqrt{2ME}-\beta\sqrt{2ME'}\right)}_{=0 \text{ if } \alpha \neq \beta}a_\beta(E') \\
&= a_\alpha(E)^*\delta_{\alpha\beta}\delta\left(\sqrt{2ME}-\sqrt{2ME'}\right)a_\beta(E'). \quad (5.1.104)
\end{aligned}$$

The δ function appearing here has vanishing argument only for $E = E'$, so it must be expressible in terms of $\delta(E-E')$. Let us, therefore, introduce $\mathcal{E} \equiv E - E'$ for the difference,

$$\delta\left(\sqrt{2ME}-\sqrt{2ME'}\right) = \delta\left(\sqrt{2M(E'+\mathcal{E})}-\sqrt{2ME'}\right) \tag{5.1.105}$$

and note that only the immediate vicinity of $\mathcal{E} = 0$ is relevant. But there we have

$$\sqrt{2M(E'+\mathcal{E})} = \sqrt{2ME'}+\sqrt{\frac{M}{2E'}}\,\mathcal{E}, \tag{5.1.106}$$

so that

$$\begin{aligned}
\delta\left(\sqrt{2ME}-\sqrt{2ME'}\right) &= \delta\left(\sqrt{\frac{M}{2E'}}\mathcal{E}\right) = \sqrt{\frac{2E'}{M}}\delta(\mathcal{E}) \\
&= \sqrt{2E'/M}\,\delta(E-E'). \tag{5.1.107}
\end{aligned}$$

The second equality here is an illustration of

$$\delta(\lambda x) = \frac{1}{\lambda}\delta(x) \quad \text{for} \quad \lambda > 0, \tag{5.1.108}$$

an identity that is used very often. Itself, this identity is a special case of

$$\delta\left(f(x)\right) = \sum_j \left|f'(x_j)\right|^{-1}\delta(x-x_j) \tag{5.1.109}$$

where $f(x)$ is a (real) function of x that has simple zeros at the x_j, and $f'(x)$ denotes the derivative of $f(x)$.

5-6 Prove this general identity for the Dirac δ function by considering the isolated immediate vicinity of one (and thus all) zeros of $f(x)$ separately.

So we have

$$\delta_{\alpha\beta}\delta(E - E') = \langle E, \alpha | E', \beta \rangle$$
$$= a_\alpha(E)^* \delta_{\alpha\beta} \sqrt{2E'/M} \delta(E - E') a_\beta(E') \quad (5.1.110)$$

and conclude that

$$|a_\alpha(E)|^2 \sqrt{2E/M} = 1 \quad (5.1.111)$$

is needed. Since there is no point in introducing arbitrary phase factors, we just take

$$a_\alpha(E) = (2E/M)^{-\frac{1}{4}} \quad (5.1.112)$$

and get

$$\big| E, \alpha \big\rangle = \big| p = \alpha\sqrt{2ME} \big\rangle (2E/M)^{-\frac{1}{4}} \quad (5.1.113)$$

for the normalized eigenkets of the force-free Hamilton operator.

5-7 Invoke the completeness relation of the momentum states to verify that

$$\sum_{\alpha=\pm} \int_0^\infty dE \, \big| E, \alpha \big\rangle\big\langle E, \alpha \big| = 1$$

holds indeed for the energy states thus normalized.

Since the energies are degenerate and we have here two states to each energy, we can take any pairwise orthogonal superpositions of same-energy eigenstates and get other sets of eigenstates, sets that are equally good. For example, the kets

$$\big| E, \text{even} \big\rangle = \Big(\big| E, + \big\rangle + \big| E, - \big\rangle \Big) \big/ \sqrt{2},$$
$$\big| E, \text{odd} \big\rangle = \Big(\big| E, + \big\rangle - \big| E, - \big\rangle \Big) \big/ (i\sqrt{2}), \quad (5.1.114)$$

will serve the purpose just as well, inasmuch as they are orthonormal

$$\langle E, a | E', b \rangle = \delta_{ab}\delta(E - E'); \quad a, b = \text{even, odd}, \quad (5.1.115)$$

and complete

$$\int_0^\infty dE \left(\big|E, \text{even}\big\rangle\big\langle E, \text{even}\big| + \big|E, \text{odd}\big\rangle\big\langle E, \text{odd}\big| \right) = 1 . \tag{5.1.116}$$

5-8 State the position wave functions $\langle x|E,\alpha\rangle$ and $\langle x|E,a\rangle$ for $\alpha = +$ or $-$ and $a =$ even or odd, and comment on them. Why is it convenient to include "i" in the definition of $|E, \text{odd}\rangle$?

5.2   Constant force

5.2.1   *Energy eigenstates*

Now, getting to more complicated matters, let us consider motion under the influence of a constant force F, so that

$$H = \frac{1}{2M}P^2 - FX , \quad F = \text{const.} \tag{5.2.1}$$

is the Hamilton operator. Indeed, Heisenberg's equations of motion,

$$\frac{d}{dt}P = \frac{1}{i\hbar}[P,H] = F ,$$
$$\frac{d}{dt}X = \frac{1}{i\hbar}[X,H] = \frac{1}{M}P , \tag{5.2.2}$$

are those of a particle of mass M exposed to a force F. The energy eigenvalues are determined by

$$H|E\rangle = |E\rangle E . \tag{5.2.3}$$

With $P \to \dfrac{\hbar}{i}\dfrac{\partial}{\partial x}$ and $X \to x$ this is

$$\left(-\frac{\hbar^2}{2M}\frac{\partial^2}{\partial x^2} - Fx \right)\psi_E(x) = E\psi_E(x) \tag{5.2.4}$$

for the position wave function $\psi_E(x) = \langle x|E\rangle$, and with $P \to p$ and $X \to i\hbar\dfrac{\partial}{\partial p}$ we get

$$\left(\frac{p^2}{2M} - i\hbar F\frac{\partial}{\partial p} \right)\psi_E(p) = E\psi_E(p) \tag{5.2.5}$$

for the momentum wave function $\psi_E(p) = \langle p|E\rangle$.

We have here a situation in which it is easier to solve the momentum variant of the time-independent Schrödinger equation. Upon introducing an integrating factor,

$$\frac{p^2}{2M} - i\hbar F \frac{\partial}{\partial p} = e^{-\frac{i}{\hbar}\frac{p^3}{6MF}} \left(-i\hbar \frac{\partial}{\partial p}\right) e^{\frac{i}{\hbar}\frac{p^3}{6MF}}, \tag{5.2.6}$$

we have

$$-i\hbar F \frac{\partial}{\partial p}\left[e^{\frac{i}{\hbar}\frac{p^3}{6MF}} \psi_E(p)\right] = E\left[e^{\frac{i}{\hbar}\frac{p^3}{6MF}} \psi_E(p)\right] \tag{5.2.7}$$

so that

$$e^{\frac{i}{\hbar}\frac{p^3}{6MF}} \psi_E(p) = a(E) e^{\frac{i}{\hbar}\frac{Ep}{F}} \tag{5.2.8}$$

or

$$\psi_E(p) = a(E) e^{-\frac{i}{\hbar}\frac{p^3}{6MF}} e^{\frac{i}{\hbar}\frac{Ep}{F}}. \tag{5.2.9}$$

The p-independent multiplicative constant $a(E)$ is fixed, to within an irrelevant phase factor, by the normalization condition

$$\langle E|E'\rangle = \delta(E - E'), \tag{5.2.10}$$

that is

$$\int dp\, \psi_E(p)^* \psi_{E'}(p) = \delta(E - E'). \tag{5.2.11}$$

There is, by the way, no restriction on the possible values for E, all real values are eigenvalues of H. Clearly we are dealing with an overidealized, oversimplified model. Real physical Hamilton operators are always bounded from below. The idealization is that of a truly constant force, constant over the whole range of x, not just a limited, perhaps large but limited, range.

The orthonormality requirement is more explicitly

$$\begin{aligned}
\delta(E - E') &= \int dp\, a(E)^* e^{\frac{i}{\hbar}\frac{p^3}{6MF}} e^{-\frac{i}{\hbar}\frac{Ep}{F}} a(E') e^{-\frac{i}{\hbar}\frac{p^3}{6MF}} e^{\frac{i}{\hbar}\frac{E'p}{F}} \\
&= a(E)^* a(E') \int dp\, e^{\frac{i}{\hbar}\left(\frac{E'}{F} - \frac{E}{F}\right)p} \\
&= a(E)^* a(E')\, 2\pi\hbar\, \delta\left(\frac{1}{F}(E' - E)\right) \\
&= 2\pi\hbar |F| |a(E)|^2 \delta(E' - E). \tag{5.2.12}
\end{aligned}$$

Opting for the simplest phase convention, $a(E) > 0$, we thus have

$$a(E) = \frac{1}{\sqrt{2\pi\hbar|F|}} \tag{5.2.13}$$

and arrive at

$$\psi_E(p) = \frac{1}{\sqrt{2\pi\hbar|F|}} \, \mathrm{e}^{-\frac{\mathrm{i}}{\hbar}\frac{p^3}{6MF}} \, \mathrm{e}^{\frac{\mathrm{i}}{\hbar}\frac{Ep}{F}} \, . \tag{5.2.14}$$

The position wave function is then available by Fourier transformation,

$$\psi_E(x) = \langle x|E \rangle = \int \mathrm{d}p \, \langle x|p \rangle \langle p|E \rangle = \int \mathrm{d}p \, \frac{\mathrm{e}^{\mathrm{i}xp/\hbar}}{\sqrt{2\pi\hbar}} \psi_E(p)$$

$$= \frac{1}{\sqrt{|F|}} \int \frac{\mathrm{d}p}{2\pi\hbar} \, \mathrm{e}^{\mathrm{i}(x + E/F)p/\hbar} \mathrm{e}^{-\frac{\mathrm{i}}{\hbar}\frac{p^3}{6MF}} \, . \tag{5.2.15}$$

5-9   The Airy function (Sir George B. Airy)

$$\mathrm{Ai}(\xi) = \int_{-\infty}^{\infty} \frac{\mathrm{d}\kappa}{2\pi} \, \mathrm{e}^{-\mathrm{i}\xi\kappa} \, \mathrm{e}^{-\frac{\mathrm{i}}{3}\kappa^3}$$

appears here. Express $\psi_E(x)$ in terms of it.

5.2.2   *Limit of no force*

We shall not take time to explore the properties of $\psi_E(x)$ in detail, but wish to see how the force-free wave functions of Section 5.1.5 emerge in the limit $F \to 0$. This limit is, in fact, a bit delicate because the physical situations of $F > 0$ and $F < 0$ are quite different. In the first case, all solutions will refer to particles moving downhill to the right, whereas $F < 0$ has downhill

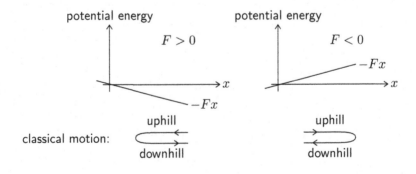

motion to the left with a clear discontinuity at $F = 0$. Indeed, $\psi_E(p)$ is singular for $F = 0$, with a singularity of the δ function kind. Therefore, it will be technically a bit easier to study the $F \to 0$ limit at the position wave function

$$\psi_E(x) = \frac{1}{\sqrt{|F|}} \int \frac{dp}{2\pi\hbar} \, e^{iW(p)/\hbar} \qquad (5.2.16)$$

with

$$W(p) = \left(x + \frac{E}{F}\right)p - \frac{p^3}{6MF}. \qquad (5.2.17)$$

The phase of the exponential $W(p)/\hbar$ changes very rapidly as a function of p when $F \cong 0$, except at the *points of stationary phase* where $\dfrac{\partial W}{\partial p} = 0$. Close to these points the otherwise rapidly oscillating $e^{iW/\hbar}$ factor is actually constant, so that p values in the vicinity contribute in phase. In other words, there is constructive interference near points of stationary phase, and destructive interference anywhere else.

The points of stationary phase are the solutions of the equation

$$\frac{\partial W}{\partial p}(p) = 0 \quad \text{or} \quad \frac{p^2}{2MF} = x + \frac{E}{F}, \qquad (5.2.18)$$

that is

$$p = \pm\bar{p}, \quad \bar{p} = \sqrt{2ME + 2MFx}. \qquad (5.2.19)$$

Near the p values of stationary phase we have

$$W(p) = W\left(\pm\bar{p} + (p \mp \bar{p})\right) \cong W(\pm\bar{p}) + \frac{1}{2}\frac{\partial^2 W}{\partial p^2}(\pm\bar{p})(p \mp \bar{p})^2 \qquad (5.2.20)$$

terminating the Taylor expansion about $p = \pm\bar{p}$ at the quadratic term. We have

$$W(\pm\bar{p}) = \pm\left(x + \frac{E}{F} - \frac{\bar{p}^2}{6MF}\right)\bar{p} = \pm\frac{\bar{p}^3}{3MF} = \pm W(\bar{p}) \qquad (5.2.21)$$

and

$$\frac{\partial^2 W}{\partial p^2}(\pm\bar{p}) = \mp\frac{\bar{p}}{MF}. \qquad (5.2.22)$$

In the p integral for $\psi_E(x)$ we thus have two major contributions for $p \cong +\bar{p}$ and $p \cong -\bar{p}$, which we wish to pick out for small F values, eventually taking the limit $F \to 0$. But then we are restricted to $E > 0$ because otherwise

$$\overline{p} \xrightarrow[F \to 0]{} \sqrt{2ME} \tag{5.2.23}$$

turns imaginary and $e^{iW(p)/\hbar}$ becomes an exponentially decreasing quantity. This mathematical detail reminds us that we have only $E > 0$ for force-free motion ($F = 0$) but arbitrary E for $F \neq 0$, however small F is.

Now, for $E > 0$, we have contributions for $p \cong \overline{p}$ and $p \cong -\overline{p}$. It is convenient to parameterize the integrals as $p = \overline{p} + q$ and $p = -\overline{p} + q$, respectively, so that

$$
\begin{aligned}
\psi_E(x) &\cong \frac{1}{\sqrt{|F|}} \int \frac{dq}{2\pi\hbar} \left[e^{i\left(W(\overline{p}) - \frac{\overline{p}}{2MF}q^2\right)/\hbar} + e^{-i\left(W(\overline{p}) - \frac{\overline{p}}{2MF}q^2\right)/\hbar} \right] \\
&= \frac{1}{\sqrt{|F|}} \frac{1}{2\pi\hbar} \left[e^{iW(\overline{p})} \sqrt{\frac{\pi}{+i\frac{\overline{p}}{2\hbar MF}}} + e^{-iW(\overline{p})} \sqrt{\frac{\pi}{-i\frac{\overline{p}}{2\hbar MF}}} \right] \\
&= \frac{1}{\sqrt{2\pi\hbar}} \left[\sqrt{\text{sgn}(F)\frac{M}{i\overline{p}}}\, e^{iW(\overline{p})} + \sqrt{\text{sgn}(F)\frac{iM}{\overline{p}}}\, e^{-iW(\overline{p})} \right]. \tag{5.2.24}
\end{aligned}
$$

Here,

$$
\begin{aligned}
W(\overline{p}) &= \frac{\overline{p}^3}{3MF} = \frac{1}{3MF}\sqrt{2ME + 2MFx}^3 \\
&= \frac{1}{3MF}\sqrt{2ME}^3\left(1 + \frac{3}{2}\frac{Fx}{E} + O(F^2)\right) \\
&\cong \frac{(2ME)^{-\frac{3}{2}}}{3MF} + \sqrt{2ME}\,x, \tag{5.2.25}
\end{aligned}
$$

keeping only the two leading terms for small F. Also

$$\overline{p}/M \to \sqrt{2E/M} \quad \text{for} \quad F \to 0, \tag{5.2.26}$$

so that

$$
\begin{aligned}
\psi_E(x) &\cong \frac{1}{\sqrt{2\pi\hbar}} (2E/M)^{-\frac{1}{4}} \left[\sqrt{\frac{1}{i}\,\text{sgn}(F)}\, e^{i\phi}\, e^{i\sqrt{2ME}x/\hbar} \right. \\
&\qquad \left. + \sqrt{i\,\text{sgn}(F)}\, e^{-i\phi}\, e^{-i\sqrt{2ME}x/\hbar} \right] \tag{5.2.27}
\end{aligned}
$$

with $\phi = \frac{1}{3MF}(2ME)^{\frac{3}{2}}$, which is an x independent phase factor.

Noting that

$$\frac{1}{\sqrt{2\pi\hbar}}\, e^{\pm i\sqrt{2ME}x/\hbar} = \langle x|p = \pm\sqrt{2ME}\rangle$$

$$= \langle x|E, \pm\rangle (2E/M)^{\frac{1}{4}} \qquad (5.2.28)$$

with $|E, \alpha\rangle$ from (5.1.113), we thus have

$$\underbrace{|E\rangle}_{\text{force } F} \cong \underbrace{|E, +\rangle}_{F=0}\, e^{i\phi}\sqrt{\frac{1}{i}\operatorname{sgn}(F)} + \underbrace{|E, -\rangle}_{F=0}\, e^{-i\phi}\sqrt{i\operatorname{sgn}(F)} \qquad (5.2.29)$$

as the $F \cong 0$ approximation. We recognize the $F = 0$ energy eigenstates, multiplied by F dependent, largely irrelevant phase factors.

5.3   Harmonic oscillator

5.3.1   *Energy eigenstates: Power-series method*

After the force-free motion, $V(X) = 0$, the motion guided by a constant force, $V(X) = -FX$, we now come to the *harmonic oscillator*, $V(X) = \frac{1}{2}M\omega^2 X^2$, for which the Hamilton operator is

$$H = \frac{1}{2M}P^2 + \frac{1}{2}M\omega^2 X^2 . \qquad (5.3.1)$$

The Heisenberg equations of motion,

$$\frac{\mathrm{d}}{\mathrm{d}t}P = \frac{1}{i\hbar}[P, H] = -\frac{\partial}{\partial X}H = -M\omega^2 X ,$$

$$\frac{\mathrm{d}}{\mathrm{d}t}X = \frac{1}{i\hbar}[X, H] = -\frac{\partial}{\partial P}H = \frac{1}{M}P , \qquad (5.3.2)$$

can be iterated to produce

$$\left(\frac{\mathrm{d}}{\mathrm{d}t}\right)^2 X = \frac{1}{M}\frac{\mathrm{d}}{\mathrm{d}t}P = -\omega^2 X , \qquad (5.3.3)$$

which is the second-order equation of motion — Newton's "acceleration = force/mass" — of a harmonic oscillator, indeed.

The time-independent Schrödinger equation for position wave functions $\psi_E(x) = \langle x|E\rangle$ reads

$$\left(-\frac{\hbar^2}{2M}\frac{\partial^2}{\partial x^2} + \frac{1}{2}M\omega^2 x^2\right)\psi_E(x) = E\psi_E(x) . \qquad (5.3.4)$$

Since the oscillator is characterized by its natural (circular) frequency ω, it is expedient to measure the energy as a multiple of $\hbar\omega$, and the parameterization

$$E = \hbar\omega\left(n + \frac{1}{2}\right) \tag{5.3.5}$$

will turn out to be particularly useful. Thus, the equation for eigenvalue E of H will turn into an equation for eigenvalue n of $H/(\hbar\omega) - \frac{1}{2}$,

$$\left(-\frac{\hbar^2}{2M\omega}\frac{\partial^2}{\partial x^2} + \frac{1}{2}\frac{M\omega}{\hbar}x^2 - \frac{1}{2}\right)\psi_n = n\psi_n \tag{5.3.6}$$

with $\psi_n \equiv \psi_E$ for $E = \hbar\omega(n + \frac{1}{2})$. The combination of factors

$$\frac{M\omega}{\hbar}x^2 = q^2 \tag{5.3.7}$$

is dimensionless ($M\omega x$ has the metrical dimension of momentum), so that

$$q = \sqrt{\frac{M\omega}{\hbar}}\,x \tag{5.3.8}$$

is the distance x expressed in the natural length unit $\sqrt{\hbar/(M\omega)}$ of the oscillator. Then

$$\frac{\hbar}{M\omega}\frac{\partial^2}{\partial x^2} = \frac{\partial^2}{\partial q^2} \tag{5.3.9}$$

and we regard ψ_n as a function of q:

$$\psi_n(q) \propto \psi_E(x) \quad \text{for} \quad E = \hbar\omega\left(n + \frac{1}{2}\right) \quad \text{and} \quad x = \frac{\hbar}{M\omega}q. \tag{5.3.10}$$

We do not write $\psi_n(q) = \psi_E(x)$ because we may find it convenient, at some later stage, to normalize $\psi_n(q)$ differently than $\psi_E(x)$.

At this stage we have

$$\left[-\frac{d^2}{dq^2} + q^2 - (2n + 1)\right]\psi_n(q) = 0. \tag{5.3.11}$$

Following a method that was systematized by Erwin Schrödinger himself, we first consider the range of very large q,

$$q \gg 1: \quad \frac{d^2}{dq^2}\psi_n(q) \cong q^2\psi_n(q) \tag{5.3.12}$$

and note that

$$\psi_n(q) = e^{\frac{1}{2}q^2} \quad \text{or} \quad e^{-\frac{1}{2}q^2} \tag{5.3.13}$$

would solve these equations (for large q values). Eventually, we shall need to normalize the solution wherefore we must insist on a finite integral of its square,

$$\int dq \, |\psi_n(q)|^2 < \infty. \tag{5.3.14}$$

It follows that $e^{+\frac{1}{2}q^2}$ is not an option, and we are led to the *ansatz*

$$\psi_n(q) = \chi_n(q) \, e^{-\frac{1}{2}q^2} \tag{5.3.15}$$

that is: we put aside the large-q exponential and regard $\chi_n(q)$ as the function to be determined.

Since

$$\frac{d}{dq}\psi_n = e^{-\frac{1}{2}q^2}\left(\frac{d}{dq} - q\right)\chi_n,$$

$$\left(\frac{d}{dq}\right)^2 \psi_n = e^{-\frac{1}{2}q^2}\left(\frac{d^2}{dq^2} - 2q\frac{d}{dq} + q^2 - 1\right)\chi_n$$

$$= [q^2 - (2n+1)]\psi_n$$

$$= e^{-\frac{1}{2}q^2}(q^2 - 2n - 1)\chi_n, \tag{5.3.16}$$

we get

$$\left(\frac{d^2}{dq^2} - 2q\frac{d}{dq} + 2n\right)\chi_n = 0 \tag{5.3.17}$$

for the differential equation obeyed by $\chi_n(q)$. It contains, of course, the unknown eigenvalue $n = E/(\hbar\omega) - \frac{1}{2}$.

We write $\chi_n(q)$ as a power series in q,

$$\chi_n(q) = \sum_{k=0}^{\infty} \frac{1}{k!} a_k^{(n)} q^k, \tag{5.3.18}$$

so that

$$\frac{d^2}{dq^2}\chi_n(q) = \sum_{k=2}^{\infty} \frac{1}{(k-2)!} a_k^{(n)} q^{k-2}$$

$$= \sum_{k=0}^{\infty} \frac{1}{k!} a_{k+2}^{(n)} q^k \qquad (5.3.19)$$

and

$$q\frac{d}{dq}\chi_n(q) = \sum_{k=0}^{\infty} \frac{1}{k!} k a_k^{(n)} q^k . \qquad (5.3.20)$$

Together they give us, as the coefficient of q^k,

$$a_{k+2}^{(n)} - 2k a_k^{(n)} + 2n a_k^{(n)} = 0 , \qquad (5.3.21)$$

which is a simple recurrence relation for the coefficients $a_k^{(n)}$,

$$a_{k+2}^{(n)} = 2(k - n) a_k^{(n)} . \qquad (5.3.22)$$

It is actually two recurrence relations — one for even k, one for odd k. Thus, given the values of $a_0^{(n)}$ and $a_1^{(n)}$, we can work out all $a_k^{(n)}$, and so determine $\chi_n(q)$, and then have a solution of the differential equation. This solution depends on the value chosen for n, which is the looked-for eigenvalue that we should determine first.

To this end, let us assume we have found an eigenvalue n. Then

$$k = 2j \text{ even}: \ a_{2(j+1)}^{(n)} = 4\left(j - \frac{n}{2}\right) a_{2j}^{(n)} ,$$

$$k = 2j + 1 \text{ odd}: \ a_{2(j+1)+1}^{(n)} = 4\left(j - \frac{n-1}{2}\right) a_{2j+1}^{(n)} , \qquad (5.3.23)$$

for $j = 0, 1, 2, \dots$. For large k, that is large j, these are approximately

$$a_{2(j+1)}^{(n)} \cong 4j a_{2j}^{(n)} \cong 4(j+1) a_{2j}^{(n)} ,$$

$$a_{2(j+1)+1}^{(n)} \cong 4j a_{2j+1}^{(n)} \cong 4(j+1) a_{2j+1}^{(n)} , \qquad (5.3.24)$$

or

$$a_{2j}^{(n)} \propto 4^j j! \quad \text{and} \quad a_{2j+1}^{(n)} \propto 4^j j! . \qquad (5.3.25)$$

So the even terms together would amount to roughly

$$\propto \sum_j \frac{1}{(2j)!} 4^j j! q^{2j} \qquad (5.3.26)$$

and the odd terms roughly to something similar. We focus on large j here, so that James Stirling's approximation

$$j! \cong \sqrt{2\pi j}\, j^j\, e^{-j} \quad \text{(for } j \gg 1) \tag{5.3.27}$$

applies, giving us

$$\frac{4^j j!}{(2j)!} \cong \frac{4^j \sqrt{2\pi j}\, j^j\, e^{-j}}{\sqrt{4\pi j}\, (2j)^{2j}\, e^{-2j}} = \frac{1}{\sqrt{2}} j^{-j}\, e^j \cong \frac{\sqrt{\pi j}}{j!}. \tag{5.3.28}$$

Accordingly,

$$\sum_j \frac{4^j j!}{(2j)!} q^{2j} \cong \sum_j \frac{\sqrt{\pi j}}{j!} q^{2j} \tag{5.3.29}$$

grows like e^{q^2} (in fact even a bit faster because of the $\sqrt{j}$ in the numerator), so that the large-j behavior of $\chi_n(q)$ is dominated by e^{q^2}, and that of $\psi_n(q)$,

$$\psi_n(q) = \chi_n(q)\, e^{-\frac{1}{2}q^2} \sim e^{\frac{1}{2}q^2}, \tag{5.3.30}$$

is then of the form that we had to discard earlier. The only way out of this dilemma is that eigenvalue n must be an integer itself so that $k - n$ is zero for some k and then the recurrence relation stops. In other words, there are no $a_k^{(n)} \neq 0$ for arbitrary large k values, and therefore we do not have this problem.

In summary, then, we must have

$$n = \text{even integer}, \quad a_n^{(n)} \neq 0, \quad a_{2j}^{(n)} = 0 \text{ for } 2j > n,$$

$$a_{2j+1}^{(n)} = 0 \text{ for all } j;$$

$$\text{or} \quad n = \text{odd integer}, \quad a_n^{(n)} \neq 0, \quad a_{2j+1}^{(n)} = 0 \text{ for } 2j+1 > n,$$

$$a_{2j}^{(n)} = 0 \text{ for all } j. \tag{5.3.31}$$

That is: $\chi_n(q)$ is a polynomial of degree n that is even in q if n is even, odd in q if n is odd. If we put, by matter of convention,

$$a_n^{(n)} = 2^n \tag{5.3.32}$$

then the $\chi_n(q)$ are the *Hermite polynomials* $H_n(q)$, their name honoring Charles Hermite. They can be defined, as we did, by their differential equation,

$$\left(\frac{d^2}{dq^2} - 2q\frac{d}{dq} + 2n \right) H_n(q) = 0, \tag{5.3.33}$$

in conjunction with requiring that the highest power of q is

$$H_n(q) = (2q)^n + \{\text{powers } q^{n-2}, q^{n-4}, \ldots\}. \tag{5.3.34}$$

The symmetry

$$H_n(-q) = (-1)^n H_n(q) \tag{5.3.35}$$

then follows.

The energy eigenvalues of the harmonic oscillator are thus

$$E_n = \hbar\omega\left(n + \frac{1}{2}\right) \quad \text{with} \quad n = 0, 1, 2 \ldots$$

$$= \frac{1}{2}\hbar\omega, \frac{3}{2}\hbar\omega, \frac{5}{2}\hbar\omega, \ldots \tag{5.3.36}$$

and the corresponding wave functions are of the form

$$\psi_n(x) = a_n H_n(q)\, e^{-\frac{1}{2}q^2} \quad \text{with} \quad q = \sqrt{\frac{M\omega}{\hbar}}\, x \tag{5.3.37}$$

and a normalization factor a_n that is undetermined as yet. We shall get it in passing later; see (5.3.75) and (5.3.82) below. As a consequence of the symmetry property (5.3.35), the oscillator eigenstates are odd states when n is odd and even states when n is even.

Note that the ground-state energy to

$$H = \frac{1}{2M}P^2 + \frac{1}{2}M\omega^2 X^2 \tag{5.3.38}$$

is $\frac{1}{2}\hbar\omega$ (obtained for $n = 0$, of course). But we know that it is always possible to add an arbitrary constant to H without changing anything of relevance. Thus

$$H = \frac{1}{2M}P^2 + \frac{1}{2}M\omega^2 X^2 - \frac{1}{2}\hbar\omega \quad \text{with} \quad E_n = n\hbar\omega \tag{5.3.39}$$

would be just as good, and it is quite often more convenient to use this alternative Hamilton operator, for which the ground-state energy is $E_0 = 0$.

5.3.2 Energy eigenstates: Ladder-operator approach

Since we measure energies in units of $\hbar\omega$ we can exhibit this factor in H itself by writing

$$H = \hbar\omega\left(\frac{M\omega}{2\hbar}X^2 + \frac{1}{2\hbar M\omega}P^2\right). \tag{5.3.40}$$

This looks like the absolute square of a complex number with real part $\propto X$ and imaginary part $\propto P$, but we are dealing with noncommuting operators here and, therefore, the absolute square of

$$A = \sqrt{\frac{M\omega}{2\hbar}} X + \mathrm{i} \frac{1}{\sqrt{2\hbar M\omega}} P \qquad (5.3.41)$$

requires multiplication with the adjoint operator,

$$A^\dagger = \sqrt{\frac{M\omega}{2\hbar}} X - \mathrm{i} \frac{1}{\sqrt{2\hbar M\omega}} P, \qquad (5.3.42)$$

and the multiplication order matters. See

$$\left.\begin{matrix} AA^\dagger \\ A^\dagger A \end{matrix}\right\} = \frac{M\omega}{2\hbar} X^2 + \frac{1}{2\hbar M\omega} P^2 \mp \mathrm{i} \frac{1}{2\hbar} \underbrace{(XP - PX)}_{= \mathrm{i}\hbar}$$

$$= \frac{1}{\hbar\omega} H \pm \frac{1}{2}. \qquad (5.3.43)$$

Accordingly, we have the commutation relation

$$[A, A^\dagger] = 1 \qquad (5.3.44)$$

and can express the Hamilton operator in various ways, including these three versions:

$$H = \hbar\omega\left(A^\dagger A + \frac{1}{2}\right) = \hbar\omega\left(AA^\dagger - \frac{1}{2}\right)$$

$$= \frac{1}{2}\hbar\omega\left(A^\dagger A + AA^\dagger\right), \qquad (5.3.45)$$

the first of which is most commonly used.

The energy eigenstates $|n\rangle$,

$$H|n\rangle = |n\rangle\hbar\omega\left(n + \frac{1}{2}\right), \qquad (5.3.46)$$

are therefore eigenstates of $A^\dagger A$,

$$A^\dagger A|n\rangle = |n\rangle n \quad \text{with} \quad n = 0, 1, 2, \ldots . \qquad (5.3.47)$$

In this context, it is common to refer to the states $|n\rangle$ as the *Fock states*, named after Vladimir A. Fock. In particular, we have

$$\langle n|A^\dagger A|n\rangle = n, \qquad (5.3.48)$$

which tells us that the ket $A|n\rangle$ has length $\sqrt{n}$. For $n = 0$ it follows that

$$A|0\rangle = 0 \,. \tag{5.3.49}$$

Therefore, the ground-state wave function $\psi_0(x) = \langle x|0\rangle$ obeys

$$0 = \langle x|A|0\rangle = \left(\sqrt{\frac{M\omega}{2\hbar}}\, x + \mathrm{i} \frac{1}{\sqrt{2\hbar M\omega}} \frac{\hbar}{\mathrm{i}} \frac{\partial}{\partial x} \right) \psi_0(x) \tag{5.3.50}$$

or

$$0 = \left(\frac{\partial}{\partial x} + \frac{M\omega}{\hbar} x \right) \psi_0(x) \,. \tag{5.3.51}$$

This is (4.8.4) with

$$\frac{1}{2(\delta X)^2} = \frac{M\omega}{\hbar} \quad \text{or} \quad \delta X = \sqrt{\frac{\hbar}{2M\omega}} \tag{5.3.52}$$

and, therefore, $\psi_0(x)$ is the minimum-uncertainty wave function of (4.8.10),

$$\psi_0(x) = \left(\frac{M\omega}{\pi\hbar} \right)^{\frac{1}{4}} \mathrm{e}^{-\frac{1}{2}\frac{M\omega}{\hbar}x^2} \,. \tag{5.3.53}$$

The corresponding momentum wave function $\psi_0(p) = \langle p|0\rangle$ is available in (4.8.15) with

$$\delta P = \frac{\hbar/2}{\delta X} = \sqrt{\frac{1}{2}\hbar M\omega} \,, \tag{5.3.54}$$

so that

$$\psi_0(p) = (\pi\hbar M\omega)^{-\frac{1}{4}} \mathrm{e}^{-\frac{1}{2}\frac{1}{\hbar M\omega}p^2} \,. \tag{5.3.55}$$

Now notice this:

$$A^\dagger A A^\dagger = A^\dagger \big(\underbrace{AA^\dagger - A^\dagger A}_{=\,1} + A^\dagger A \big) = A^\dagger \left(A^\dagger A + 1 \right) \,. \tag{5.3.56}$$

Apply it to $|n\rangle$,

$$A^\dagger A A^\dagger |n\rangle = A^\dagger \left(A^\dagger A + 1 \right)|n\rangle = A^\dagger |n\rangle (n+1) \,, \tag{5.3.57}$$

and learn that $A^\dagger |n\rangle$ is the eigenket of $A^\dagger A$ with eigenvalue $n+1$. Therefore, it must be proportional to the eigenket $|n+1\rangle$,

$$A^\dagger |n\rangle \propto |n+1\rangle \,, \tag{5.3.58}$$

and

$$\left(\langle n|A\right)\left(A^\dagger|n\rangle\right) = \langle n|AA^\dagger|n\rangle$$
$$= \langle n|(A^\dagger A + 1)|n\rangle = n + 1 \tag{5.3.59}$$

implies that

$$A^\dagger|n\rangle = |n+1\rangle\sqrt{n+1} \tag{5.3.60}$$

up to an arbitrary phase factor, which we put equal to 1, thereby adopting the standard conventions.

The adjoint statement is

$$\langle n|A = \sqrt{n+1}\langle n+1|. \tag{5.3.61}$$

We act with $A^\dagger$ on the right,

$$\langle n|AA^\dagger = \langle n|(A^\dagger A + 1) = (n+1)\langle n|$$
$$= \sqrt{n+1}\langle n+1|A^\dagger, \tag{5.3.62}$$

and conclude that

$$\langle n+1|A^\dagger = \sqrt{n+1}\langle n|, \tag{5.3.63}$$

to which

$$A|n+1\rangle = |n\rangle\sqrt{n+1} \tag{5.3.64}$$

is the adjoint. To ease memorization, we collect all these statements here:

$$A^\dagger|n\rangle = |n+1\rangle\sqrt{n+1}, \qquad \langle n|A = \sqrt{n+1}\langle n+1|,$$
$$A|n\rangle = |n-1\rangle\sqrt{n}, \qquad \langle n|A^\dagger = \sqrt{n}\langle n-1|. \tag{5.3.65}$$

These relations show how A and $A^\dagger$ raise or lower the quantum number n by 1. In view of this property they are called *ladder operators* — you go up or down the n ladder by one rung upon applying $A^\dagger$ or A to the bra or ket in question. It is customary to speak of the "raising operator $A^\dagger$", and the "lowering operator A", which is their action on kets $|n\rangle$, but keep in mind that the roles of raising and lowering operators are reversed for bras $\langle n|$.

This raising and lowering action by itself can be used to determine the eigenvalues of $A^\dagger A$, and thus of the Hamilton operator H, in a purely algebraic manner without resorting to a direct solution of the time-independent

Schrödinger equation (5.3.4) or (5.3.11). So, let us pretend that we do not know as yet the eigenvalues of $A^\dagger A$, but agree to write

$$A^\dagger A|n\rangle = |n\rangle n\,, \quad n = ?\,. \tag{5.3.66}$$

The ladder operator property is an application of the commutation relation $[A, A^\dagger] = 1$, so it is surely true that $A^\dagger|n\rangle = |n + 1\rangle\sqrt{n + 1}$ and so forth. This is to say that if n is eigenvalue of $A^\dagger A$, then also $n + 1$ is eigenvalue of $A^\dagger A$, so we can climb up the ladder. Climbing down is done by applying A, $A|n\rangle = |n - 1\rangle\sqrt{n}$, but that cannot go on forever because we know from (5.3.48) that the squared length of $A|n\rangle$ equals n, which therefore cannot be negative.

So, when climbing down the ladder there must be a last rung, a bottom rung, for which the application of A does not give another eigenket of $A^\dagger A$. This is only possible if $n = 0$ marks the bottom rung because $A|n = 0\rangle = 0$. It follows that $n = 0$ is the smallest eigenvalue of $A^\dagger A$, and then $n = 1, 2, 3, \ldots$ are the other eigenvalues, obtained by moving up on the ladder. Together we have

$$n = 0, 1, 2, 3, \ldots\,, \tag{5.3.67}$$

and we get this result without the harder work in Section 5.3.1, where we extracted this information out of the Schrödinger eigenvalue equation for $A^\dagger A = H/\hbar\omega - \frac{1}{2}$.

In fact, we are further rewarded by the algebraic method because we are told that

$$|n\rangle = \frac{A^\dagger}{\sqrt{n}}|n - 1\rangle = \frac{A^\dagger}{\sqrt{n}}\frac{A^\dagger}{\sqrt{n - 1}}|n - 2\rangle$$

$$= \cdots = \frac{(A^\dagger)^n}{\sqrt{n!}}|0\rangle \tag{5.3.68}$$

that is: we get the eigenkets of $A^\dagger A$ together with their normalization. Beginning with

$$\psi_0(x) = \langle x|0\rangle = \left(\frac{M\omega}{\pi\hbar}\right)^{\frac{1}{4}} e^{-\frac{1}{2}\frac{M\omega}{\hbar}x^2} = \left(\frac{M\omega}{\pi\hbar}\right)^{\frac{1}{4}} e^{-\frac{1}{2}q^2} \tag{5.3.69}$$

we can thus get the wave functions for all eigenstates of the harmonic oscillator by repeated application of $A^\dagger$,

$$\psi_n(x) = \langle x|n\rangle = \frac{1}{\sqrt{n!}}\langle x|(A^\dagger)^n|n\rangle \tag{5.3.70}$$

with

$$\langle x|A^{\dagger} = \langle x| \left(\sqrt{\frac{M\omega}{2\hbar}} X - \mathrm{i}\frac{1}{\sqrt{2M\hbar\omega}} P \right)$$
$$= \left(\sqrt{\frac{M\omega}{2\hbar}} x - \mathrm{i}\frac{1}{\sqrt{2M\hbar\omega}} \frac{\hbar}{\mathrm{i}} \frac{\partial}{\partial x} \right) \langle x|$$
$$= \frac{1}{\sqrt{2}} \left(q - \frac{\partial}{\partial q} \right) \langle x| . \tag{5.3.71}$$

Accordingly,

$$\psi_n(x) = \left(\frac{M\omega}{\pi\hbar} \right)^{\frac{1}{4}} \frac{1}{\sqrt{n!}} \left[\frac{1}{\sqrt{2}} \left(q - \frac{\mathrm{d}}{\mathrm{d}q} \right) \right]^n \mathrm{e}^{-\frac{1}{2}q^2}$$
$$= \left(\frac{M\omega}{\pi\hbar} \right)^{\frac{1}{4}} \frac{1}{\sqrt{2^n n!}} \left(q - \frac{\mathrm{d}}{\mathrm{d}q} \right)^n \mathrm{e}^{-\frac{1}{2}q^2} . \tag{5.3.72}$$

This becomes simpler once we note that

$$q - \frac{\mathrm{d}}{\mathrm{d}q} = \mathrm{e}^{\frac{1}{2}q^2} \left(-\frac{\mathrm{d}}{\mathrm{d}q} \right) \mathrm{e}^{-\frac{1}{2}q^2} \tag{5.3.73}$$

and therefore

$$\left(q - \frac{\mathrm{d}}{\mathrm{d}q} \right)^n = \mathrm{e}^{\frac{1}{2}q^2} \left(-\frac{\mathrm{d}}{\mathrm{d}q} \right)^n \mathrm{e}^{-\frac{1}{2}q^2} . \tag{5.3.74}$$

We have thus arrived at

$$\psi_n(x) = \left(\frac{M\omega}{\pi\hbar} \right)^{\frac{1}{4}} \frac{1}{\sqrt{2^n n!}} \mathrm{e}^{\frac{1}{2}q^2} \left(-\frac{\mathrm{d}}{\mathrm{d}q} \right)^n \mathrm{e}^{-q^2} , \tag{5.3.75}$$

the position wave function of the nth Fock state, which we compare with the earlier result (5.3.37) and infer that

$$\mathrm{H}_n(q) = \mathrm{e}^{q^2} \left(-\frac{\mathrm{d}}{\mathrm{d}q} \right)^n \mathrm{e}^{-q^2} . \tag{5.3.76}$$

To begin with, there could be room here for a multiplicative constant, but in fact there is no extra factor because the normalization of $\mathrm{H}_n(q)$ in accordance with (5.3.34) is exactly what the above formula gives, because the leading power is obtained by applying each $-\dfrac{\mathrm{d}}{\mathrm{d}q}$ to e^{-q^2}, each application supplying one factor of $2q$.

5.3.3 *Hermite polynomials*

There are very many mathematical identities, recurrence relations and others, about Hermite polynomials, and quite a number of equivalent definitions for them. We have met two: The differential equation (5.3.11), and the *Rodrigues formula* (5.3.76), which exemplifies a type of equation that is named after Benjamin O. Rodrigues. Arguably the simplest and also the most important formula is the *generating function*

$$e^{2tq - t^2} = \sum_{n=0}^{\infty} \frac{t^n}{n!} H_n(q) \tag{5.3.77}$$

from which all statements about $H_n(q)$ can be derived easily.

This generating function can be found by going backward, see:

$$\sum_{n=0}^{\infty} \frac{t^n}{n!} H_n(q) = e^{q^2} \sum_{n=0}^{\infty} \frac{t^n}{n!} \left(\frac{d}{dq} \right)^n e^{-q^2}$$

$$= e^{q^2} e^{-t\frac{d}{dq}} e^{-q^2} = e^{q^2} e^{-(q-t)^2}$$

$$= e^{2qt - t^2} . \tag{5.3.78}$$

As a simple application we consider the orthogonality relation of the Hermite polynomials,

$$\int_{-\infty}^{\infty} dq \, e^{-q^2} H_n(q) H_m(q) = \boxed{?}_{mn} . \tag{5.3.79}$$

We deal with these numbers as a set: multiply by $\dfrac{t^n}{n!} \dfrac{s^m}{m!}$ and sum over n and m,

$$\sum_{n,m=0}^{\infty} \frac{t^n}{n!} \frac{s^m}{m!} \boxed{?}_{nm} = \int_{-\infty}^{\infty} dq \, e^{-q^2} e^{2qt - t^2} e^{2sq - s^2}$$

$$= \int dq \, e^{-[q - (t+s)]^2} e^{2ts} = \sqrt{\pi} \, e^{2ts}$$

$$= \sqrt{\pi} \sum_{n=0}^{\infty} \frac{(2ts)^n}{n!}$$

$$= \sqrt{\pi} \sum_{n,m=0}^{\infty} \frac{2^n t^n s^m}{n!} \delta_{nm} , \tag{5.3.80}$$

so that $\boxed{?}_{nm} = 2^n n! \sqrt{\pi} \delta_{nm}$ and

$$\int_{-\infty}^{\infty} dq\, e^{-q^2} H_n(q) H_m(q) = \delta_{nm} 2^n n! \sqrt{\pi}. \qquad (5.3.81)$$

In the physical context of eigenstates of the harmonic oscillator, this is essentially the orthonormality of the Fock state kets $|n\rangle$, explicitly specified by their position wave functions

$$\psi_n(x) = \left(\frac{M\omega}{\pi\hbar}\right)^{\frac{1}{4}} (2^n n!)^{-\frac{1}{2}} H_n(q)\, e^{-\frac{1}{2}q^2} \quad \text{with} \quad q = \sqrt{\frac{M\omega}{\hbar}} x. \qquad (5.3.82)$$

We check this:

$$\begin{aligned}
\langle n|m\rangle &= \int dx\, \langle n|x\rangle\langle x|m\rangle = \int dx\, \psi_n(x)^* \psi_n(x) \\
&= \left(\frac{M\omega}{\pi\hbar}\right)^{\frac{1}{2}} (2^{n+m} n! m!)^{-\frac{1}{2}} \underbrace{\int dx}_{= \sqrt{\hbar/(M\omega)}\, dq}\, e^{-q^2} H_n(q) H_m(q) \\
&= \left(2^{n+m} n! m! \pi\right)^{-\frac{1}{2}} \delta_{nm} 2^n n! \sqrt{\pi} \\
&= \delta_{nm}, \qquad\qquad\qquad\qquad\qquad\qquad\qquad\qquad (5.3.83)
\end{aligned}$$

indeed.

5.3.4 Infinite matrices

When using the oscillator eigenkets $|0\rangle, |1\rangle, \ldots$ as the basis kets and their bras as the basis bras, an arbitrary operator is represented by an *infinite* matrix,

$$\begin{aligned}
Z &= \sum_{n=0}^{\infty} |n\rangle\langle n| Z \sum_{m=0}^{\infty} |m\rangle\langle m| = \sum_{n,m=0}^{\infty} |n\rangle\langle n|Z|m\rangle\langle m| \\
&= \left(|0\rangle, |1\rangle, \ldots\right) \begin{pmatrix} \langle 0|Z|0\rangle & \langle 0|Z|1\rangle & \cdots \\ \langle 1|Z|0\rangle & \langle 1|Z|1\rangle & \cdots \\ \vdots & \vdots & \ddots \end{pmatrix} \begin{pmatrix} \langle 0| \\ \langle 1| \\ \vdots \end{pmatrix} \\
&\cong \begin{pmatrix} \langle 0|Z|0\rangle & \langle 0|Z|1\rangle & \cdots \\ \langle 1|Z|0\rangle & \langle 1|Z|1\rangle & \cdots \\ \vdots & \vdots & \ddots \end{pmatrix}, \qquad\qquad\qquad (5.3.84)
\end{aligned}$$

and kets are represented by infinitely long columns,

$$| \rangle = (|0\rangle, |1\rangle, \ldots) \begin{pmatrix} \langle 0| \; \rangle \\ \langle 1| \; \rangle \\ \vdots \end{pmatrix} \cong \begin{pmatrix} \langle 0| \; \rangle \\ \langle 1| \; \rangle \\ \vdots \end{pmatrix} = \begin{pmatrix} \psi_0 \\ \psi_1 \\ \vdots \end{pmatrix}, \qquad (5.3.85)$$

and bras by infinitely long rows. Matrix multiplication, or the multiplication of a matrix with a column, then involves an infinite summation, a series, and the convergence properties offer the usual treat of possible pitfalls, easily avoided as a rule by a bit of careful attention to such details.

As illustrating examples let us find the infinite matrices for the position operator X and the momentum operator P. We recall that

$$A = \sqrt{\frac{M\omega}{2\hbar}} X + i \frac{1}{\sqrt{2\hbar M\omega}} P,$$

$$A^\dagger = \sqrt{\frac{M\omega}{2\hbar}} X - i \frac{1}{\sqrt{2\hbar M\omega}} P, \qquad (5.3.86)$$

and note that

$$X = \sqrt{\frac{\hbar}{2M\omega}} (A^\dagger + A),$$

$$P = \sqrt{\frac{\hbar M\omega}{2}} (iA^\dagger - iA). \qquad (5.3.87)$$

Therefore we have

$$\langle n|X|m \rangle = \sqrt{\frac{\hbar}{2M\omega}} \langle n|(A^\dagger + A)|m \rangle$$

$$= \sqrt{\frac{\hbar}{2M\omega}} \left(\langle n|m+1\rangle \sqrt{m+1} + \sqrt{n+1}\langle n+1|m \rangle \right)$$

$$= \sqrt{\frac{\hbar}{2M\omega}} \left(\delta_{n,m+1}\sqrt{m+1} + \sqrt{n+1}\delta_{n+1,m} \right)$$

$$= \sqrt{\frac{\hbar}{2M\omega}} \left(\sqrt{n}\delta_{n,m+1} + \sqrt{m}\delta_{n+1,m} \right) \qquad (5.3.88)$$

and likewise

$$\langle n|P|m \rangle = \sqrt{\frac{\hbar}{2M\omega}} \left(i\sqrt{n}\delta_{n,m+1} - i\sqrt{m}\delta_{n+1,m} \right). \qquad (5.3.89)$$

Quite explicitly the matrices are

$$X \cong \sqrt{\frac{\hbar}{2M\omega}} \begin{pmatrix} 0 & \sqrt{1} & 0 & \cdots & \cdots & \cdots \\ \sqrt{1} & 0 & \sqrt{2} & 0 & \cdots & \cdots \\ 0 & \sqrt{2} & 0 & \sqrt{3} & 0 & \cdots \\ \vdots & 0 & \sqrt{3} & 0 & \sqrt{4} & \cdots \\ \vdots & \vdots & \vdots & \vdots & \vdots & \ddots \end{pmatrix} \tag{5.3.90}$$

and

$$P \cong \sqrt{\frac{\hbar M\omega}{2}} \begin{pmatrix} 0 & -i\sqrt{1} & 0 & & \cdots & & \cdots & \cdots \\ i\sqrt{1} & 0 & -i\sqrt{2} & 0 & & \cdots & \cdots \\ 0 & i\sqrt{2} & 0 & -i\sqrt{3} & 0 & \cdots \\ \vdots & 0 & i\sqrt{3} & 0 & -i\sqrt{4} & \cdots \\ \vdots & \vdots & \vdots & \vdots & \vdots & \ddots \end{pmatrix}, \tag{5.3.91}$$

and it is an easy task to verify that

$$XP - PX = [X, P] \cong i\hbar \underbrace{\begin{pmatrix} 1 & 0 & \cdots & \cdots & \cdots \\ 0 & 1 & 0 & \cdots & \cdots \\ \vdots & 0 & 1 & 0 & \cdots \\ \vdots & \vdots & \vdots & \ddots & \ddots \end{pmatrix}}_{\text{infinite identity matrix}}, \tag{5.3.92}$$

as it should be.

We know from the discussion in Section 5.3.2 that the ground state ($n = 0$) of the harmonic oscillator is a minimum-uncertainty state. For arbitrary n, we first note that

$$\langle X \rangle_n = \sqrt{\frac{\hbar}{2M\omega}} \langle n | (A^\dagger + A) | n \rangle = 0 \,,$$

$$\langle P \rangle_n = \sqrt{\frac{\hbar M\omega}{2}} \langle n | (iA^\dagger - iA) | n \rangle = 0 \,, \tag{5.3.93}$$

and further we have

$$\left\langle X^2 \right\rangle_n = \frac{\hbar}{2M\omega} \langle n| \left(\underset{\underset{0}{\downarrow}}{A^{\dagger 2}} + \underset{\underset{n}{\downarrow}}{A^{\dagger}A} + \underset{\underset{n+1}{\downarrow}}{AA^{\dagger}} + \underset{\underset{0}{\downarrow}}{A^2} \right) |n\rangle$$

$$= \frac{\hbar}{M\omega} \left(n + \frac{1}{2} \right) \qquad (5.3.94)$$

and

$$\left\langle P^2 \right\rangle_n = \frac{\hbar M\omega}{2} \langle n| \left(iA^{\dagger} - iA \right)^2 |n\rangle$$

$$= \frac{\hbar M\omega}{2} \langle n| \left(\underset{\underset{0}{\downarrow}}{-A^{\dagger 2}} + \underset{\underset{n}{\downarrow}}{A^{\dagger}A} + \underset{\underset{n+1}{\downarrow}}{AA^{\dagger}} - \underset{\underset{0}{\downarrow}}{A^2} \right) |n\rangle$$

$$= \hbar M\omega \left(n + \frac{1}{2} \right). \qquad (5.3.95)$$

Then

$$\delta X_n = \sqrt{\frac{\hbar}{M\omega}} \sqrt{n + \frac{1}{2}},$$

$$\delta P_n = \sqrt{\hbar M\omega} \sqrt{n + \frac{1}{2}}, \qquad (5.3.96)$$

for the position spread and momentum spread in the nth oscillator state. Their products

$$\delta X_n \, \delta P_n = \hbar \left(n + \frac{1}{2} \right) = \frac{1}{2}\hbar, \ \frac{3}{2}\hbar, \ \frac{5}{2}\hbar, \ \dots \qquad (5.3.97)$$

always exceed the Heisenberg lower bound of $\frac{1}{2}\hbar$ for $n > 0$.

5-10 The expectation values in (5.3.93) are the diagonal elements of the matrices in (5.3.90) and (5.3.91). Determine the expectation values in (5.3.94) and (5.3.95) as the diagonal elements of the squares of these matrices.

5.4   Delta potential

5.4.1   *Bound state*

On a couple of occasions — see, for instance, Exercises 4-2 and 4-7 on pages 116 and 125, respectively — we used the wave function

$$\psi(x) = \sqrt{\kappa}\, e^{-\kappa|x|}, \quad \kappa > 0 \tag{5.4.1}$$

as an example. Its derivative

$$\frac{\partial}{\partial x}\psi(x) = -\sqrt{\kappa^3}\, \mathrm{sgn}\,(x)\, e^{-\kappa|x|} \tag{5.4.2}$$

involves the derivative of $|x|$,

$$\frac{\mathrm{d}}{\mathrm{d}x}|x| = \mathrm{sgn}\,(x) = \begin{cases} +1 & \text{for} \quad x > 0, \\ -1 & \text{for} \quad x < 0, \end{cases} \tag{5.4.3}$$

the *sign function*. And the second derivative of $\psi(x)$ needs the derivative thereof,

$$\frac{\mathrm{d}}{\mathrm{d}x}\,\mathrm{sgn}\,(x) = 2\delta(x) \tag{5.4.4}$$

jumps by $+2$ at $x = 0$

so that

$$\left(\frac{\partial}{\partial x}\right)^2 \psi(x) = \sqrt{\kappa^5}\, e^{-\kappa|x|} - 2\sqrt{\kappa^3}\delta(x)\, e^{-\kappa|x|}$$

$$= \kappa^2\psi(x) - 2\kappa\delta(x)\psi(x). \tag{5.4.5}$$

5-11   Consider the three cases $a < 0 < b$, $0 < a < b$, and $a < b < 0$ to verify that

$$\int_a^b \mathrm{d}x\, \delta(x) = \frac{1}{2}\big[\mathrm{sgn}(b) - \mathrm{sgn}(a)\big],$$

thereby demonstrating the assertion of (5.4.4), namely that $\frac{1}{2}\,\mathrm{sgn}(x)$ is an antiderivative of the Dirac δ function.

We compare (5.4.5) with the Schrödinger eigenvalue equation for a Hamilton operator with potential energy $V(X)$,

$$H = \frac{1}{2M} P^2 + V(X), \tag{5.4.6}$$

that is

$$-\frac{\hbar^2}{2M} \left(\frac{\partial}{\partial x}\right)^2 \psi_E(x) + V(x)\psi_E(x) = E\psi_E(x) \tag{5.4.7}$$

or

$$\left(\frac{\partial}{\partial x}\right)^2 \psi_E(x) = -\frac{2ME}{\hbar^2} \psi_E(x) + \frac{2M}{\hbar^2} V(x)\psi_E(x) \tag{5.4.8}$$

and observe that $\psi(x) = \sqrt{\kappa}\, e^{-\kappa|x|}$ solves this equation for

$$\kappa^2 = -\frac{2ME}{\hbar^2} \quad \text{and} \quad -2\kappa\delta(x) = \frac{2M}{\hbar^2} V(x) \tag{5.4.9}$$

or

$$V(x) = -\frac{\hbar^2\kappa}{M}\delta(x) \quad \text{and} \quad E = -\frac{(\hbar\kappa)^2}{2M}. \tag{5.4.10}$$

The Hamilton operator in question is therefore given by

$$H = \frac{1}{2M} P^2 - \frac{\hbar^2\kappa}{M}\delta(X). \tag{5.4.11}$$

The Dirac δ function of position operator X that appears here has a rather simple meaning, inasmuch as

$$\delta(X) = \int dx\, |x\rangle\delta(x)\langle x|$$
$$= \left(|x\rangle\langle x|\right)_{x=0} = |x=0\rangle\langle x=0|\,; \tag{5.4.12}$$

it is the projector to the state for eigenvalue $x = 0$ of operator X. Geometrically speaking, the potential energy

$$V(x) = -\frac{\hbar^2 \kappa}{M} \delta(x)$$

(5.4.13)

approximates the more physical situation of a very narrow, very deep potential well — or, put differently, a very strong force of very short range.

The modeling of the more realistic physical situation by the mathematical idealization of a δ potential is justifiable if all relevant distances are large compared to the range of the potential. In particular the distance over which the wave function of interest changes significantly must be large in this sense. If this is the case, the simple $\delta(x)$ potential catches the essence of the strong short-range force well and we use it as a good physical approximation to the more complicated real thing.

Suppose we had not known the solution to the eigenvalue equation to begin with, but have to find $|E\rangle$ such that

$$\left[\frac{1}{2M}P^2 - \frac{\hbar^2\kappa}{M}\delta(X)\right]|E\rangle = |E\rangle E\,.$$

(5.4.14)

How could we go about it? We could, of course, turn this into the differential equation for $\langle x|E\rangle$ and then solve it as it comes. It is, however, simpler and also instructive to look for the momentum wave function $\psi_E(p) = \langle p|E\rangle$ first.

Since $\delta(X) = |x=0\rangle\langle x=0|$ we have

$$\langle p|\delta(X)|E\rangle = \langle p|x=0\rangle\langle x=0|E\rangle$$

$$= \left[\frac{e^{-ipx/\hbar}}{\sqrt{2\pi\hbar}}\right]_{x=0} \psi_E(x=0)$$

$$= \frac{1}{\sqrt{2\pi\hbar}}\psi_E(x=0)\,,$$

(5.4.15)

where $\psi_E(x=0)$ is a number to be determined. Then (5.4.14) turns into

$$\frac{1}{2M}p^2\psi_E(p) - \frac{\hbar^2\kappa}{M}\frac{\psi_E(x=0)}{\sqrt{2\pi\hbar}} = E\psi_E(p). \qquad (5.4.16)$$

Upon writing

$$E = -\frac{q^2}{2M} \quad \text{with} \quad q > 0 \qquad (5.4.17)$$

this appears as

$$(p^2 + q^2)\psi_E(p) = 2\hbar^2\kappa\frac{\psi_E(x=0)}{\sqrt{2\pi\hbar}} \equiv C. \qquad (5.4.18)$$

The solution

$$\psi_E(p) = \frac{C}{p^2 + q^2} \qquad (5.4.19)$$

will give us also the position wave function by Fourier transformation

$$\psi_E(x) = \int dp \, \langle x|p \rangle \underbrace{\langle p|E \rangle}_{=\psi_E(p)}$$

$$= \int dp \frac{e^{ixp/\hbar}}{\sqrt{2\pi\hbar}} \frac{C}{p^2 + q^2}. \qquad (5.4.20)$$

But before we evaluate this integral for arbitrary x, let us look at the consistency condition that obtains for $x = 0$:

$$\psi_E(x=0) = \int dp \frac{C}{\sqrt{2\pi\hbar}} \frac{1}{p^2 + q^2}$$

$$= \frac{C}{\sqrt{2\pi\hbar}} \underbrace{\int \frac{dp}{p^2 + q^2}}_{=\pi/q} = \frac{\hbar\kappa}{q}\psi_E(x=0). \qquad (5.4.21)$$

This tells us that $q = \hbar\kappa$ is the only possible value for q and that, therefore, the only negative eigenvalue of H is $E = -(\hbar\kappa)^2/(2M)$.

We thus have

$$\psi_E(p) = \frac{C}{p^2 + (\hbar\kappa)^2} \qquad (5.4.22)$$

and can determine the modulus of C, and thus of $\psi_E(x = 0)$, from the normalization $\langle E|E\rangle = 1$,

$$1 = \int dp\, |\psi_E(p)|^2 = |C|^2 \underbrace{\int \frac{dp}{(p^2 + q^2)^2}}_{= \frac{\pi}{2}/q^3} \quad \text{with} \quad q = \hbar\kappa \qquad (5.4.23)$$

so that

$$|C| = \sqrt{\frac{2}{\pi}}(\hbar\kappa)^{\frac{3}{2}} \quad \text{and} \quad |\psi_E(x = 0)| = \sqrt{\kappa}\,, \qquad (5.4.24)$$

consistent, of course, with (5.4.1). Opting for the phase choice specified by $\psi_E(x = 0) > 0$ we get

$$\psi_E(x = 0) = \sqrt{\kappa} \qquad (5.4.25)$$

and

$$\psi_E(p) = \frac{\sqrt{2/\pi}(\hbar\kappa)^{\frac{3}{2}}}{p^2 + (\hbar\kappa)^2}\,. \qquad (5.4.26)$$

We return to the Fourier integral in (5.4.20) and evaluate it by a contour integration and an application of the residue method. You may skip this part if you do not know enough complex analysis.

The integral in question,

$$\psi_E(x) = \frac{C}{\sqrt{2\pi\hbar}} \int dp\, \frac{e^{ixp/\hbar}}{p^2 + q^2} \qquad (5.4.27)$$

defines a function that is even in x because the odd part of

$$e^{ixp/\hbar} = \underbrace{\cos(xp/\hbar)}_{\text{even,}} + \underbrace{i\sin(xp/\hbar)}_{\text{odd in }p\text{ and }x} \qquad (5.4.28)$$

does not contribute to the p integral. We can therefore write

$$\psi_E(x) = \frac{C}{\sqrt{2\pi\hbar}} \int dp\, \frac{e^{i|x|p/\hbar}}{p^2 + q^2}$$

$$= \frac{C}{\sqrt{2\pi\hbar}} \frac{1}{2iq} \int dp \left(\frac{e^{i|x|p/\hbar}}{p - iq} - \frac{e^{i|x|p/\hbar}}{p + iq} \right). \qquad (5.4.29)$$

The latter form exhibits the poles at $p = iq$ and $p = -iq$,

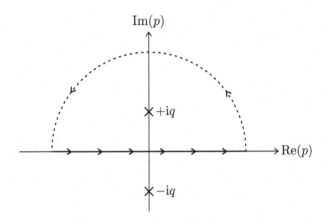

and we close the integration path by a very large half circle in the upper plane where $\mathrm{Im}(p) > 0$ and $e^{i|x|p/\hbar}$ is exponentially small. The p integral is therefore given by the residue at $p = +iq$,

$$\psi_E(x) = \frac{C}{\sqrt{2\pi\hbar}} \frac{1}{2iq} 2\pi i \underbrace{e^{i|x|iq/\hbar}}_{\text{residue at } p = +iq}$$

$$= \sqrt{\frac{\pi}{2\hbar}} \frac{C}{q} e^{-|x|q/\hbar} \qquad (5.4.30)$$

or, with $q = \hbar\kappa$ and $C = \sqrt{2\hbar\kappa/\pi}\,\hbar\kappa$, as required by (5.4.21) and (5.4.26),

$$\psi_E(x) = \sqrt{\kappa}\, e^{-\kappa|x|}, \qquad (5.4.31)$$

indeed. Not surprisingly, we have returned to our starting point (5.4.1).

5.4.2 *Scattering states*

Having thus established that there is one and only one negative-energy eigenvalue of

$$H = \frac{1}{2M} P^2 - \frac{\hbar^2\kappa}{M} \delta(X) \qquad (5.4.32)$$

with energy $E = -(\hbar\kappa)^2/(2M)$, all other eigenstates must have positive energy, $E = (\hbar k)^2/(2M)$ with $k > 0$. For the force-free motion we had even states, $\psi(x) \propto \cos(kx)$, and odd states, $\psi(x) \propto \sin(kx)$. The latter wave functions vanish at $x = 0$ and, therefore, these remain solutions of

$$\left[-\frac{\hbar^2}{2M} \frac{\partial^2}{\partial x^2} - \frac{\hbar^2 \kappa}{M} \delta(x) \right] \psi(x) = \frac{(\hbar k)^2}{2M} \psi(x) \,. \qquad (5.4.33)$$

So, odd solutions are

$$\psi_{\text{odd}}(x) = \sin(kx) \,, \qquad (5.4.34)$$

where we do not care about the proper normalization (for the sake of simplicity only — we could care if we wanted to).

The even solutions will be of the form

$$\psi_{\text{even}}(x) = \cos(k|x| - \alpha) \,, \qquad (5.4.35)$$

with a phase shift α that we need to determine. It is clear that this is the generic form of the even wave functions, because for $x \neq 0$ we must have some linear combinations of $\cos(kx)$ and $\sin(kx)$, or equivalently one of them with a shifted argument, such as $\cos(kx + \text{const})$. Upon enforcing $\psi(-x) = \psi(x)$, the stated form is implied.

We need the second derivative of $\psi_{\text{even}}(x)$,

$$\begin{aligned}
\left(\frac{\partial}{\partial x} \right)^2 \psi_{\text{even}}(x) &= \frac{\mathrm{d}}{\mathrm{d}x} \left[-k \operatorname{sgn}(x) \sin(k|x| - \alpha) \right] \\
&= -k^2 \cos(k|x| - \alpha) - 2k\delta(x) \sin(k|x| - \alpha) \\
&= -k^2 \psi_{\text{even}}(x) + 2k\delta(x) \sin \alpha \,. \qquad (5.4.36)
\end{aligned}$$

We insert it into the eigenvalue differential equation (5.4.33),

$$\frac{\partial^2}{\partial x^2} \psi_{\text{even}}(x) + 2\kappa\, \delta(x) \underbrace{\psi_{\text{even}}(x)}_{\rightarrow\, \psi_{\text{even}}(0)\, =\, \cos \alpha} = -k^2 \psi_{\text{even}}(x) \,, \qquad (5.4.37)$$

and get

$$k \sin \alpha + \kappa \cos \alpha = 0 \qquad (5.4.38)$$

or

$$\tan \alpha = -\frac{\kappa}{k} \,, \qquad (5.4.39)$$

so that

$$\kappa = 0: \quad \alpha = 0\,,$$

$$k \gg \kappa: \quad \alpha \cong -\frac{\kappa}{k}\,,$$

$$k \ll \kappa: \quad \alpha \cong -\frac{\pi}{2}\,. \tag{5.4.40}$$

In particular the $\alpha = 0$ value for $\kappa = 0$ is expected, because we must recover the force-free solution $\psi_{\mathrm{even}}(x) = \cos(kx) = \cos(k|x|)$ for $\kappa = 0$.

For $E > 0$, we thus have

$$\text{odd solutions} \quad \psi_{\mathrm{odd}}(x) = \sin(kx) \tag{5.4.41}$$

and

$$\text{even solutions} \quad \psi_{\mathrm{even}}(x) = \cos(k|x| - \alpha) \tag{5.4.42}$$

with $k = \sqrt{2mE/\hbar^2} > 0$ and $\alpha = -\arctan(\kappa/k)$.

We ask the question: Which linear combination of them,

$$\psi(x) = A\cos(k|x| - \alpha) + B\sin(kx) \tag{5.4.43}$$

is such that

$$\psi(x) = \mathrm{e}^{\mathrm{i}kx} \quad \text{for} \quad x > 0\,, \tag{5.4.44}$$

and what is then $\psi(x)$ for $x < 0$? We have, for $x > 0$,

$$\mathrm{e}^{\mathrm{i}kx} = A\Big(\cos(kx)\cos\alpha + \sin(kx)\sin\alpha\Big) + B\sin(kx)$$

$$= A\cos\alpha\cos(kx) + (A\sin\alpha + B)\sin(kx) \tag{5.4.45}$$

and thus need

$$A\cos\alpha = 1\,, \quad A\sin\alpha + B = \mathrm{i} \tag{5.4.46}$$

or

$$A = \frac{1}{\cos\alpha} = \sqrt{1 + (\tan\alpha)^2} = \sqrt{1 + (\kappa/k)^2}\,,$$

$$B = \mathrm{i} - A\sin\alpha = \mathrm{i} - \tan\alpha = \mathrm{i} + \frac{\kappa}{k}\,. \tag{5.4.47}$$

For $x < 0$, we obtain

$$\begin{aligned}
\psi(x) &= A\cos(kx + \alpha) + B\sin(kx) \\
&= \frac{1}{\cos\alpha}\big(\cos(kx)\cos\alpha - \sin(kx)\sin\alpha\big) + (\mathrm{i} - \tan\alpha)\sin(kx) \\
&= \cos(kx) + (\mathrm{i} - 2\tan\alpha)\sin(kx) \qquad\qquad\qquad (5.4.48)
\end{aligned}$$

or

$$\psi(x) = (1 + \mathrm{i}\tan\alpha)\,\mathrm{e}^{\mathrm{i}kx} - \mathrm{i}\tan\alpha\,\mathrm{e}^{-\mathrm{i}kx}. \qquad\qquad (5.4.49)$$

In summary, this position wave function of an eigenstate to energy $E = (\hbar k)^2/(2M)$ is of the form

$$\psi(x) = \begin{cases} (1 + \mathrm{i}\tan\alpha)\,\mathrm{e}^{\mathrm{i}kx} - \mathrm{i}\tan\alpha\,\mathrm{e}^{-\mathrm{i}kx} & \text{for} \quad x < 0, \\ \mathrm{e}^{\mathrm{i}kx} & \text{for} \quad x > 0. \end{cases} \qquad (5.4.50)$$

Recalling that $P \to \dfrac{\hbar}{\mathrm{i}}\dfrac{\partial}{\partial x}$ for position wave functions, we have momentum $+\hbar k$ for $\mathrm{e}^{\mathrm{i}kx}$ and momentum $-\hbar k$ for $\mathrm{e}^{-\mathrm{i}kx}$, which tells us that $\mathrm{e}^{\mathrm{i}kx}$ is the amplitude for motion to the right (from negative to positive x values) and $\mathrm{e}^{-\mathrm{i}kx}$ for motion to the left:

We have here a particularly simple case of the more general situation discussed in Section 3.2 of *Perturbed Evolution*.

The wave function (5.4.50) is not normalized so that the amplitude factors that multiply the plane wave exponential factors do not have absolute meaning, but only relative meaning:

part of $\psi(x)$	relative amplitude	normalized
incoming, $\mathrm{e}^{\mathrm{i}kx}$, $x < 0$	$1 + \mathrm{i}\tan\alpha$	1
transmitted, $\mathrm{e}^{\mathrm{i}kx}$, $x > 0$	1	$\dfrac{1}{1 + \mathrm{i}\tan\alpha}$
reflected, $\mathrm{e}^{-\mathrm{i}kx}$, $x < 0$	$-\mathrm{i}\tan\alpha$	$\dfrac{-\mathrm{i}\tan\alpha}{1 + \mathrm{i}\tan\alpha}$

In the last column, we have the amplitudes that result from normalizing the incoming part to unit amplitude, so that the transmitted and reflected amplitudes are then relative to the incoming one. They are

$$\text{transmission:} \quad \frac{1}{1 + i\tan\alpha} = e^{-i\alpha}\cos\alpha,$$

$$\text{reflection:} \quad \frac{-i\tan\alpha}{1 + i\tan\alpha} = -ie^{-i\alpha}\sin\alpha, \qquad (5.4.51)$$

and we find the respective probabilities by squaring,

$$\text{prob(transmission)} = \left| e^{-i\alpha}\cos\alpha \right|^2 = (\cos\alpha)^2 = \frac{k^2}{\kappa^2 + k^2},$$

$$\text{prob(reflection)} = \left| -ie^{-i\alpha}\sin\alpha \right|^2 = (\sin\alpha)^2 = \frac{\kappa^2}{\kappa^2 + k^2}. \quad (5.4.52)$$

Note that the qualitative aspects are as could have been expected. If the potential is of large relative strength, $\kappa \gg k$, the reflection probability is very large; if it is relatively weak, $\kappa \ll k$, then reflection is unlikely and transmission highly probable. In the force-free limit, $\kappa = 0$, there is of course no reflection at all.

Note also that the sign of κ has no bearing on these probabilities. For either sign in

$$H = \frac{1}{2M}P^2 \mp \frac{\hbar^2\kappa}{M}\delta(X) \qquad (5.4.53)$$

we get the same probabilities for reflection and transmission. This is not really a generic feature, but it is typical: Usually attractive and repulsive potentials are difficult to distinguish on their *scattering* properties alone.

This last sentence introduces some terminology. The states for $E > 0$, where we have a *continuum* of energy eigenvalues, are called *scattering states* because they refer to physical situations of the kind just discussed: an incoming probability amplitude wave combined with transmitted and reflected ones. By contrast, the states to *discrete* energy eigenvalues, in the present example only the one for $E = -(\hbar\kappa)^2/(2M)$, are *bound states*, their position wave functions decrease rapidly (usually exponentially) when you move away from the area where the potential energy $V(x)$ is relevant.

5.5   Square-well potential

5.5.1   *Bound states*

Harkening back to the discussion on page 183, let us now take a look at the situation of an attractive square-well potential of finite width a and finite depth V_0:

$$V(x) = \begin{cases} 0 & \text{for} \quad |x| > a/2, \\ -V_0 < 0 & \text{for} \quad |x| < a/2. \end{cases}$$

$$(5.5.1)$$

We focus on *bound states*, $E < 0$,

$$H|E\rangle = |E\rangle E \qquad (5.5.2)$$

of the Hamilton operator

$$H = \frac{1}{2M}P^2 + V(X). \qquad (5.5.3)$$

Since $\langle H \rangle = E$, then, and $\langle P^2 \rangle > 0$ while $0 > \langle V \rangle > -V_0$, we have

$$0 > E > -V_0 \qquad (5.5.4)$$

or $E < 0$ and $E + V_0 > 0$.

The Schrödinger eigenvalue equation for $\psi_E(x) = \langle x|E \rangle$ is

$$\left[-\frac{\hbar^2}{2M}\frac{\partial^2}{\partial x^2} + V(x) \right]\psi_E(x) = E\psi_E(x), \qquad (5.5.5)$$

that is

$$\frac{\partial^2}{\partial x^2}\psi_E(x) = -\frac{2ME}{\hbar^2}\psi_E(x) \quad \text{for} \quad |x| > a/2,$$

$$\frac{\partial^2}{\partial x^2}\psi_E(x) = -\frac{2M(V_0 + E)}{\hbar^2}\psi_E(x) \quad \text{for} \quad |x| < a/2. \qquad (5.5.6)$$

Upon introducing the energy parameters κ and k by means of

$$E = -\frac{(\hbar\kappa)^2}{2M}, \quad \kappa > 0,$$

$$E + V_0 = +\frac{(\hbar k)^2}{2M}, \quad k > 0, \tag{5.5.7}$$

we have

$$\frac{\partial^2}{\partial x^2}\psi_E(x) = \begin{cases} +\kappa^2\psi_E(x) & \text{for} \quad |x| > a/2, \\ -k^2\psi_E(x) & \text{for} \quad |x| < a/2. \end{cases} \tag{5.5.8}$$

Therefore, *even* solutions are of the form

$$\psi_{\text{even}}(x) = \begin{cases} A\,e^{-\kappa|x|} & \text{for} \quad |x| > a/2, \\ B\cos(kx) & \text{for} \quad |x| < a/2, \end{cases} \tag{5.5.9}$$

and *odd* solutions are of the form

$$\psi_{\text{odd}}(x) = \begin{cases} A\,\text{sgn}\,(x)\,e^{-\kappa|x|} & \text{for} \quad |x| > a/2, \\ B\sin(x) & \text{for} \quad |x| < a/2, \end{cases} \tag{5.5.10}$$

where A and B are to be chosen and that the respective $\psi(x)$s are continuous, and also their derivatives,

$$\frac{\partial}{\partial x}\psi_{\text{even}}(\kappa) = \begin{cases} -A\kappa\,\text{sgn}\,(x)\,e^{-\kappa|x|} & \text{for} \quad |x| > a/2, \\ -Bk\sin(kx) & \text{for} \quad |x| < a/2, \end{cases}$$

$$\frac{\partial}{\partial x}\psi_{\text{odd}}(x) = \begin{cases} -A\kappa\,e^{-\kappa|x|} & \text{for} \quad |x| > a/2, \\ Bk\cos(x) & \text{for} \quad |x| < a/2. \end{cases} \tag{5.5.11}$$

The continuity of $\psi(x)$ at $x = a/2$ requires

$$\text{even:} \quad A\,e^{-\kappa a/2} = B\cos(ka/2),$$

$$\text{odd:} \quad A\,e^{-\kappa a/2} = B\sin(ka/2), \tag{5.5.12}$$

and the continuity of $\partial\psi/\partial x$ at $x = a/2$ requires

$$\text{even:} \quad -\kappa A\,e^{-\kappa a/2} = -kB\sin(ka/2),$$

$$\text{odd:} \quad -\kappa A\,e^{-\kappa a/2} = kB\cos(ka/2). \tag{5.5.13}$$

Together,

$$\text{even:} \quad \kappa = k \tan(ka/2),$$
$$\text{odd:} \quad \kappa = -k \cot(ka/2). \tag{5.5.14}$$

It is convenient to write $\vartheta \equiv ka/2$ and then

$$\left(\frac{\kappa a}{2}\right)^2 = \underbrace{\frac{2MV_0}{\hbar^2}\left(\frac{a}{2}\right)^2}_{\equiv\,\theta^2} + \underbrace{\frac{2ME}{\hbar^2}\left(\frac{a}{2}\right)^2}_{=\,-(ka/2)^2\,=\,-\vartheta^2} = \theta^2 - \vartheta^2. \tag{5.5.15}$$

Thus, we now parameterize the looked-for energy eigenvalues by parameter ϑ in accordance with

$$E = -V_0 + \frac{(\hbar k)^2}{2M} = -V_0 + \frac{(2\hbar/a)^2}{2M}\vartheta^2, \quad \vartheta > 0 \tag{5.5.16}$$

and summarize the essential details of the potential energy in the corresponding parameter θ defined by

$$V_0 = \frac{(2\hbar/a)^2}{2M}\theta^2, \quad \theta > 0. \tag{5.5.17}$$

Equations (5.5.14) then turn into

$$\text{even:} \quad \sqrt{\theta^2 - \vartheta^2} = \vartheta \tan \vartheta,$$
$$\text{odd:} \quad \sqrt{\theta^2 - \vartheta^2} = -\vartheta \cot \vartheta. \tag{5.5.18}$$

Their solutions, for ϑ, tell us the negative energy eigenvalues of the Hamilton operator in (5.5.3), that is its *bound-state energies*.

These are transcendental equations that we cannot solve analytically, but it is easy to get solutions numerically. We can, however, establish a number of facts without knowing all these numerical details. For this purpose we use a graphical representation of the two sides of the equation.

For the *even* case, it is sketched as

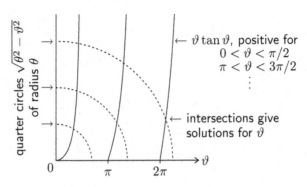

and we see that there is

$$1 \text{ solution for } \theta < \pi,$$
$$2 \text{ solutions for } \pi < \theta < 2\pi,$$
$$3 \text{ solutions for } 2\pi < \theta < 3\pi,$$
$$\vdots$$
$$n \text{ solutions for } (n-1)\pi < \theta < n\pi. \tag{5.5.19}$$

For *odd* wave functions we have this picture

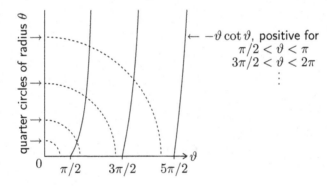

and here we read off that there is

$$\text{no solution for } \theta < \pi/2,$$
$$1 \text{ solution for } \pi/2 < \theta < 3\pi/2,$$
$$2 \text{ solutions for } 3\pi/2 < \theta < 5\pi/2,$$
$$\vdots$$
$$n \text{ solutions for } (2n-1)\pi/2 < \theta < (2n+1)\pi/2. \tag{5.5.20}$$

We get ϑ values for even solutions in the intervals $0 \cdots \frac{\pi}{2}$, $\pi \cdots \frac{3\pi}{2}$, $2\pi \cdots \frac{5\pi}{2}$, ..., and ϑ values for odd solutions in the intervals $\frac{\pi}{2} \cdots \pi$, $\frac{3\pi}{2} \cdots 2\pi$, $\frac{5\pi}{2} \cdots 3\pi$, ... Accordingly, more and more bound states become available as θ increases, beginning with just one even bound state, then adding a first odd one, then a second even one, then a second odd one, and so forth. Altogether there are n bound states if

$$n - 1 < \frac{2}{\pi}\theta < n \qquad (5.5.21)$$

with either the same number of even and odd states (if $n = 2, 4, 6, \ldots$) or one more even state than there are odd states (if $n = 1, 3, 5, \ldots$).

Many of the details of this little calculation about a finite square-well potential are naturally pertinent to this potential only, such as the transcendental equations (5.5.18) for the values of ϑ and thus the precise energy eigenvalues. Other features are, however, generic. Any even potential, $V(x) = V(-x)$, that supports bound states (that is an attractive potential) will have a lowest-energy state, a ground state, with an even wave function (no nodes), possibly an odd first excited state (one node), then perhaps a second, even excited state (two nodes) and so forth. Potentials that are not even can also lead to bound states, with zero nodes for the ground state, one node for the first excited state, two nodes for the second excited state .... In fact, it is customary to label the bound states by their number of nodes. In the present example of the finite square well, we would then write

$$\psi_0(x), \psi_2(x), \psi_4(x), \ldots \text{ for the even solutions,}$$
$$\psi_1(x), \psi_3(x), \psi_5(x), \ldots \text{ for the odd solutions,} \qquad (5.5.22)$$

with $\vartheta_0, \vartheta_2, \vartheta_4, \ldots$ solving the "even" equation in (5.5.18) and $\vartheta_1, \vartheta_3, \vartheta_5,$ ... solving the "odd" one, whereby

$$\frac{\pi}{2}k < \vartheta_k < \frac{\pi}{2}(k + 1) \qquad (5.5.23)$$

and $-V_0 < E_0 < E_1 < \cdots < 0$

$$\text{for} \quad E_k = -V_0 + \frac{(2\hbar/a)^2}{2M}\vartheta_k^2. \qquad (5.5.24)$$

5.5.2 *Delta potential as a limit*

Let us briefly consider the limit of the δ function potential, that is

$$V_0 \to \infty, \quad a \to 0 \quad \text{such that} \quad V_0 a \to \frac{\hbar^2 \kappa}{M} \tag{5.5.25}$$

where this κ is the one of Section 5.4. So let us write

$$V_0 a = \frac{\hbar^2 \kappa}{M} \tag{5.5.26}$$

and then

$$\theta^2 = 2M \left(\frac{a}{2\hbar} \right)^2 V_0 = \frac{1}{2} \kappa a, \tag{5.5.27}$$

leading to

$$\vartheta^2 + (\vartheta \tan \vartheta)^2 = \theta^2 = \frac{1}{2} \kappa a. \tag{5.5.28}$$

The term $(\vartheta \tan \vartheta)^2$ is $\propto \vartheta^4$ for small ϑ^2, so that

$$\vartheta^2 = \frac{1}{2} \kappa a - \left(\frac{1}{2} \kappa a \right)^2 + \cdots \tag{5.5.29}$$

for small values of κa, as they are relevant here, in the limit of $a \to 0$ with κ fixed. This ϑ is, of course, the ϑ_0 of above, and the ground-state energy is

$$
\begin{aligned}
E_0 &= -V_0 + \frac{(2\hbar/a)^2}{2M} \vartheta_0^2 \\
&= -V_0 + \frac{(2\hbar/a)^2}{2M} \left(\frac{1}{2} \kappa a - \left(\frac{1}{2} \kappa a \right)^2 + \cdots \right) \\
&= \underbrace{-V_0 + \frac{\hbar^2 \kappa}{Ma}}_{=0} - \underbrace{\frac{(\hbar \kappa)^2}{2M} + \cdots}_{\to 0 \text{ as } a \to 0} \\
&\to -\frac{(\hbar \kappa)^2}{2M} \quad \text{for} \quad a \to 0, \quad \text{as it should.}
\end{aligned}
\tag{5.5.30}
$$

Indeed, we recover the earlier result for the ground-state energy of the attractive δ potential, the only negative energy eigenvalue available in this limit.

5.5.3   Scattering states. Tunneling

As we did for the delta potential in Section 5.4.2, we could consider the scattering states for the attractive square-well potential of (5.5.1). Somewhat more interesting is, however, the repulsive square-well potential

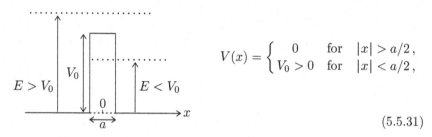

$$V(x) = \begin{cases} 0 & \text{for} \quad |x| > a/2\,, \\ V_0 > 0 & \text{for} \quad |x| < a/2\,, \end{cases}$$

$$(5.5.31)$$

where we expect a qualitative difference between high energies, $E > V_0$, and low energies, $0 < E < V_0$, because the region $-a/2 < x < a/2$ could not be accessed in classical mechanics for low energies.

Indeed, classical intuition would let us expect that the atom is surely transmitted for $E > V_0$ and surely reflected for $E < V_0$. But what does quantum mechanics say about the actual situation?

The answer is given by the solution of the time-independent Schrödinger equation (5.5.5) for $V(x)$ of (5.5.31). In analogy with (5.5.7) we introduce parameters k and κ by means of

$$E = \frac{(\hbar k)^2}{2M}\,, \quad k > 0\,,$$

$$|E - V_0| = \frac{(\hbar \kappa)^2}{2M}\,, \quad \kappa > 0\,.$$

$$(5.5.32)$$

This turns (5.5.5) into

$$\frac{\partial^2}{\partial x^2}\psi_E(x) = -k^2\psi_E(x) \quad \text{for} \quad |x| > \frac{a}{2}$$

$$(5.5.33)$$

and

$$\frac{\partial^2}{\partial x^2}\psi_E(x) = \begin{cases} +\kappa^2\psi_E(x) & \text{for} \quad |x| < \frac{a}{2} \text{ and } E < V_0\,, \\ -\kappa^2\psi_E(x) & \text{for} \quad |x| < \frac{a}{2} \text{ and } E > V_0\,. \end{cases}$$

$$(5.5.34)$$

Guided by the discussion in Section 5.4.2, we are looking for a solution with the structure of the wave function in (5.4.50), as depicted on page 189: an incoming amplitude from the left, a transmitted amplitude outgoing to

the right, and a reflected amplitude outgoing to the left. This is to say that
we use

$$\psi_E(x) = \begin{cases} e^{ikx} + r\,e^{-ikx} & \text{for} \quad x < -\dfrac{a}{2}\,, \\[2mm] t\,e^{ikx} & \text{for} \quad x > \dfrac{a}{2}\,, \end{cases} \qquad (5.5.35)$$

for the two regions where (5.5.33) applies. We thus normalize the incoming
amplitude, that is: e^{ikx} for $x < -a/2$, to unity, so that the *reflected amplitude* r and the *transmitted amplitude* t give us the respective probabilities
by squaring,

$$\text{prob(reflection)} = |r|^2\,,$$
$$\text{prob(transmission)} = |t|^2\,. \qquad (5.5.36)$$

In the region $-a/2 < x < a/2$, the distinction between low and high
energies is crucial, and in view of (5.5.34) we have

$$\psi_E(x) = \begin{cases} A\cosh(\kappa x) + B\sinh(\kappa x) & \text{for} \quad |x| < \dfrac{a}{2} \text{ and } E < V_0\,, \\[2mm] A\cos(\kappa x) + B\sin(\kappa x) & \text{for} \quad |x| < \dfrac{a}{2} \text{ and } E > V_0\,. \end{cases}$$
$$(5.5.37)$$

The four amplitude coefficients — r and t in (5.5.35) as well as A and B
in (5.5.37) — are determined by the continuity of $\psi_E(x)$ and $\dfrac{\partial}{\partial x}\psi_E(x)$ at
$x = -a/2$ and $x = +a/2$, which together are four conditions indeed.

The two continuity conditions at $x = -a/2$ are

$$\psi_E\left(-\frac{a}{2}\right) = e^{-ika/2} + r\,e^{ika/2}$$
$$= \begin{cases} A\cosh\dfrac{\kappa a}{2} - B\sinh\dfrac{\kappa a}{2} & \text{for} \quad E < V_0\,, \\[2mm] A\cos\dfrac{\kappa a}{2} - B\sin\dfrac{\kappa a}{2} & \text{for} \quad E > V_0\,, \end{cases} \qquad (5.5.38)$$

and

$$\frac{1}{\kappa}\frac{\partial\psi_E}{\partial x}\left(-\frac{a}{2}\right) = \frac{ik}{\kappa}\left(e^{-ika/2} + r\,e^{ika/2}\right)$$
$$= \begin{cases} -A\sinh\dfrac{\kappa a}{2} + B\cosh\dfrac{\kappa a}{2} & \text{for} \quad E < V_0\,, \\[2mm] A\sin\dfrac{\kappa a}{2} + B\cos\dfrac{\kappa a}{2} & \text{for} \quad E > V_0\,. \end{cases} \qquad (5.5.39)$$

And at $x = +a/2$ we have the continuity conditions

$$\psi_E\left(\frac{a}{2}\right) = t\,e^{ika/2}$$

$$= \begin{cases} A\cosh\dfrac{\kappa a}{2} + B\sinh\dfrac{\kappa a}{2} & \text{for } E < V_0, \\[2mm] A\cos\dfrac{\kappa a}{2} + B\sin\dfrac{\kappa a}{2} & \text{for } E > V_0, \end{cases} \tag{5.5.40}$$

and

$$\frac{1}{\kappa}\frac{\partial\psi_E}{\partial x}\left(\frac{a}{2}\right) = \frac{ik}{\kappa}t\,e^{ika/2}$$

$$= \begin{cases} A\sinh\dfrac{\kappa a}{2} + B\cosh\dfrac{\kappa a}{2} & \text{for } E < V_0, \\[2mm] -A\sin\dfrac{\kappa a}{2} + B\cos\dfrac{\kappa a}{2} & \text{for } E > V_0. \end{cases} \tag{5.5.41}$$

We take a look at the low-energy case of $E < V_0$, for which (5.5.38) and (5.5.39) are compactly summarized by

$$e^{ika/2}\begin{pmatrix} 1 & 1 \\[1mm] \dfrac{ik}{\kappa} & -\dfrac{ik}{\kappa} \end{pmatrix}\begin{pmatrix} e^{ika} \\[1mm] r \end{pmatrix} = \begin{pmatrix} \cosh\dfrac{\kappa a}{2} & -\sinh\dfrac{\kappa a}{2} \\[2mm] -\sinh\dfrac{\kappa a}{2} & \cosh\dfrac{\kappa a}{2} \end{pmatrix}\begin{pmatrix} A \\[1mm] B \end{pmatrix}, \tag{5.5.42}$$

and (5.5.40) and (5.5.41) by

$$e^{ika/2}t\begin{pmatrix} 1 \\[1mm] \dfrac{ik}{\kappa} \end{pmatrix} = \begin{pmatrix} \cosh\dfrac{\kappa a}{2} & \sinh\dfrac{\kappa a}{2} \\[2mm] \sinh\dfrac{\kappa a}{2} & \cosh\dfrac{\kappa a}{2} \end{pmatrix}\begin{pmatrix} A \\[1mm] B \end{pmatrix}. \tag{5.5.43}$$

Since the two 2×2 matrices on the right are inverses of each other, we can immediately solve (5.5.42) for A and B, and insert them into (5.5.43). The outcome

$$t\begin{pmatrix} 1 \\[1mm] \dfrac{ik}{\kappa} \end{pmatrix} = \begin{pmatrix} \cosh(\kappa a) & \sinh(\kappa a) \\[1mm] \sinh(\kappa a) & \cosh(\kappa a) \end{pmatrix}\begin{pmatrix} 1 & 1 \\[1mm] \dfrac{ik}{\kappa} & -\dfrac{ik}{\kappa} \end{pmatrix}\begin{pmatrix} e^{ika} \\[1mm] r \end{pmatrix} \tag{5.5.44}$$

states two ways of expressing the transmitted amplitude t in terms of the reflected amplitude r,

$$\begin{aligned} t &= (1+r)\cosh(\kappa a) + \frac{ik}{\kappa}(1-r)\sinh(\kappa a) \\ &= (1-r)\cosh(\kappa a) + \frac{\kappa}{ik}(1+r)\sinh(\kappa a), \end{aligned} \tag{5.5.45}$$

implying first

$$r = \frac{-\mathrm{i}(\kappa^2 + k^2)\sinh(\kappa a)}{2\kappa k \cosh(\kappa a) + \mathrm{i}(\kappa^2 - k^2)\sinh(\kappa a)} \qquad (5.5.46)$$

and then

$$t = \frac{2\kappa k}{2\kappa k \cosh(\kappa a) + \mathrm{i}(\kappa^2 - k^2)\sinh(\kappa a)} \, . \qquad (5.5.47)$$

If one wishes, one can now get the values of A and B, but we skip that.

As required by (5.5.36), we square r and t to get the probabilities for reflection and transmission,

$$\mathrm{prob(reflection)} = \left[1 + \left(\frac{\kappa^2 + k^2}{2\kappa k} \sinh(\kappa a) \right)^{-2} \right]^{-1} \, ,$$

$$\mathrm{prob(transmission)} = \left[1 + \left(\frac{\kappa^2 + k^2}{2\kappa k} \sinh(\kappa a) \right)^{2} \right]^{-1} \, , \qquad (5.5.48)$$

which have unit sum, as one verifies easily, a basic check of consistency. Contrary to the classical expectation stated in page 197, there is a nonzero transmission probability in the low-energy situation to which (5.5.48) applies and, accordingly, the probability of reflection is less than 100%. This phenomenon is often referred to as "tunneling through the potential barrier" or simply known as the *tunnel effect*. More than just being an entertaining prediction of quantum theory, it is a real physical phenomenon that is exploited in devices such as Leo Esaki's tunnel diode, the tunnel transistor, or Gerd Binnig's and Heinrich Rohrer's scanning tunneling microscope.

In this context, the transmission probability is known as the *tunneling probability*. It is rather small, unless the argument of the hyperbolic sine in (5.5.48) is small, but typically it is not. Consider, for example, the extreme situation of very low energy, $0 \lesssim E \ll V_0$, when $0 \lesssim k \ll \kappa \cong \sqrt{2MV_0/\hbar^2}$ and therefore

$$\mathrm{prob(transmission)} \cong \left(\frac{2k}{\kappa \sinh(\kappa a)} \right)^2 = \frac{2E/V_0}{\sinh\left(\sqrt{2MV_0}\, a/\hbar\right)^2} \, . \qquad (5.5.49)$$

Owing to the squared hyperbolic sine in the denominator, this tunneling probability is exponentially small if the barrier width a is not tiny on the scale set by the height V_0 of the potential.

5-12   Find the probabilities for transmission and reflection for high energies, $E > V_0$. Demonstrate that there is "above-barrier reflection", another phenomenon that contradicts classical intuition.

5-13   Find the transmission and reflection probabilities for $E = V_0$ both by solving (5.5.5) for this case and by considering the limits $V_0 > E \to V_0$ and $V_0 < E \to V_0$.

Index

Note: Page numbers preceded by the letters SS or PE refer to
Simple Systems and *Perturbed Evolution*, respectively.

action   133, PE37
action principle   *see* quantum action
   principle
adiabatic approximation   PE77
adiabatic population transfer
   PE78–81
adjoint   SS11, PE1
– of a bra   37, SS11, PE1
– of a column   28
– of a ket   37, SS11, PE1
– of a ket-bra   40
– of a linear combination   SS11
– of a product   40, SS19
– of an operator   40
Airy, Sir George B.   162
Airy function   162
algebraic completeness
– of complementary pairs   PE13–17
– of position and momentum
   SS32–37
amplitude
– column of $\sim$s   95, 97
– in and out $\sim$s   PE95
– incoming $\sim$ PE91
– normalized $\sim$ 190
– probability $\sim$ PE2
– – composition law for $\sim\sim$s   PE5

– probability $\sim$ 23–32, 37–39, 109,
   SS1, 13
– – time dependent $\sim\sim$ 93–97, 105
– reflected $\sim$ 190, 198
– relative $\sim$ 190
– transmitted $\sim$ 190, 198
angular momentum   SS102, PE28
– addition   PE134–137
– and harmonic oscillators   PE134
– commutation relations   PE131
– eigenstates   SS113
– – orthonormality   SS114
– eigenvalues   SS113, PE133
– intrinsic $\sim$ *see* spin
– ladder operators   SS113, PE132
– orbital $\sim$ *see* orbital angular
   momentum
– total $\sim$ PE131
– vector operator   SS102, PE118
angular velocity   89
– vector   94
as-if reality   61, SS30
atom pairs   63–68
– entangled $\sim$ 63–68
– statistical operator   64, 65
axis vector   SS128
azimuthal wave functions

- orthonormality   SS133

Baker, Henry F.   SS40
Baker–Campbell–Hausdorff relation
    SS40, 42
Bell, John S.   4, 5
Bell correlation   5, 6, 63
Bell inequality   7, 67, 68
- violated by quantum correlations
    65, 68
Bessel, Friedrich W.   PE119
Bessel functions
- spherical $\sim$ see spherical Bessel
    functions
beta function integral   PE42
Binnig, Gerd   200
Bohr, Niels H. D.   SS125, PE17,
    116, 148
Bohr energies   SS139, 166
Bohr magneton   91, PE148
Bohr radius   SS125
Bohr shells   SS139, PE177
Bohr's principle of complementarity
    see complementarity principle
Born, Max   119, PE47
Born series   PE47, 52, 84, 114
- evolution operator   PE47
- formal summation   PE52
- scattering operator   PE49
- self-repeating pattern   PE49
- transition operator   PE114
Bose, Satyendranath   PE158
Bose–Einstein statistics   PE158
bosons   PE158, 159, 164
bound states   190
- delta potential   181–186
- hard-sphere potential   PE126
- hydrogenic atoms   SS128
- square-well potential   191–195
bra   35–38, SS10–16, PE1
- adjoint of a $\sim$ 37, PE1
- analog of row-type vector   SS11
- column of $\sim$s   39, 55, 95
- eigen$\sim$ 46–50
- infinite row for $\sim$ 178
- metrical dimension   SS19

- phase arbitrariness   SS31
- physical $\sim$ SS13
- row for $\sim$ 99, 104
bra-ket   see bracket
bracket   38–41, SS12, 24
- invariance of $\sim$s   77
Brillouin, Léon   SS159, 175
Brillouin–Wigner perturbation theory
    SS159–164
Bunyakovsky, Viktor Y.   SS14

Campbell, John E.   SS40
Carlini, Francesco   SS176
cartesian coordinates   SS117
Cauchy, Augustin-Louis   SS14
Cauchy–Bunyakovsky–Schwarz
    inequality   SS14
causal link   6
causality   1, 2
- Einsteinian $\sim$ 6, 7
center-of-mass motion
- Hamilton operator   PE156
centrifugal potential   SS123, PE118,
    123
- force-free motion   SS123
classical turning point   SS173, 177,
    183
classically allowed   SS173, 174, 177,
    183, PE92
classically forbidden   SS173, 174,
    PE92
Clebsch, Rudolf F. A.   PE137
Clebsch–Gordan coefficients   PE137
- recurrence relation   PE152
closure relation   see completeness
    relation
coherence length   128
coherent states   SS88, PE68
- and Fock states   SS90–92
- completeness relation   SS88–90, 92
- momentum wave functions   SS92
- position wave functions   SS84
column   28
- adjoint of a $\sim$ 28
- eigen$\sim$ 99
- for ket   99, 104

– normalized ∼ 18, 28
– of bras 39, 55, 95
– of coordinates SS4
– of probability amplitudes 95, 97
– orthogonal ∼s 28
– two-component ∼s 18
column-type vector SS4
– analog of ket SS10
commutation relation
– angular momentum PE131
– ladder operators 171, SS84
– velocity PE142
commutator 85
– different times SS73
– position-momentum ∼ 119, SS34, 107
– product rule 86, SS35
– sum rule 85, SS35
complementarity principle
– phenomenology PE31
– technical formulation PE17
completeness relation 41, 71, 75, SS6, 15, PE3, 23, 25, 28
– coherent states SS88–90, 92
– eigenstates of Pauli operators 50
– Fock states SS88
– force-free states 157
– momentum states 120, SS16
– position states 110, SS16
– time dependence 83, SS42
constant force 160–165, PE39–41
– Hamilton operator 160, SS63, PE39
– Heisenberg equation 160, SS63, PE39
– no-force limit 162–165
– Schrödinger equation 160, 161, SS65
– spread in momentum SS64, 66
– spread in position SS64
– time transformation function SS66, 67, PE40
– uncertainty ellipse SS65
constant of motion PE146
constant restoring force SS146, 179
– ground-state energy SS148

contour integration 185, PE106
correlation
– position-momentum ∼ SS60, 64
– quantum ∼s 65
Coulomb, Charles-Augustin SS125
Coulomb potential SS125, PE100, 101
– limit of Yukawa potential PE102
Coulomb problem see hydrogenic atoms
cyclic permutation PE7, 18
– unitary operator PE7
cyclotron frequency PE144

de Broglie, Prince Louis-Victor 121, SS176, PE92
de Broglie relation 121
de Broglie wavelength 122, 128, SS176, PE92
deflection angle PE100, 161
degeneracy
– and symmetry 157, SS99, 128
– hydrogenic atoms SS128
– of eigenenergies 157
degree of freedom PE6
– composite ∼ PE17–19
– continuous ∼ PE20, 29
– polar angle PE29
– prime ∼ PE19
– radial motion PE29
delta function 111–113, 120, 123, SS13, PE23
– antiderivative 181, SS181
– Fourier representation 124, SS18, PE24
– is a distribution 111
– model for ∼ 112, 113, PE68
– more complicated argument 158, SS20
– of position operator 182
delta potential 181–190
– as a limit 196
– bound state 181–186
– ground-state energy 196
– negative eigenvalue 184
– reflection probability 190

– scattering states   186–190
– Schrödinger equation   184, 187
– transmission probability   190
delta symbol   71, 110, SS4, PE3
– general version   SS165, PE14
delta-shell potential   PE127
density matrix   see statistical
   operator
Descartes, René   SS102
determinant   47, 49
– as product of eigenvalues   PE74
determinism   1, 2
– lack of ∼   (see also
   indeterminism)   4, 8, 9, PE30
– no hidden ∼   4–8
deterministic chaos   8
detuning   PE72
dipole moment
– electric ∼   SS172
– magnetic ∼   see magnetic moment
Dirac, Paul A. M.   39, 111, SS10,
   PE23, 158
Dirac bracket   see bracket, SS12
Dirac picture   PE49
Dirac's delta function   see delta
   function
Dirac's stroke of genius   SS12
dot product   see inner product
downhill motion   162
dyadic   SS5
– matrix for ∼   SS5
– orthogonal ∼   SS9
dynamical variables   89
– time dependence   PE32
Dyson, Freeman J.   PE52
Dyson series   PE52, 84

Eckart, Carl   PE151
effective potential   SS123, PE174
eigenbra   46–50, 72, 83, PE1
– equation   46, 83, PE1
eigenenergies   98
eigenket   46–50, 72, 83, PE1
– equation   46, 83, PE1
eigenkets
– orbital angular momentum   SS115

eigenvalue   46–52, 71, 83, PE2
– of a hermitian operator   79
– of a unitary operator   78
– trace as sum of ∼s   56
eigenvalues
– orbital angular momentum   SS115
eigenvector equation   46
Einstein, Albert   6, 51, PE158
Einsteinian causality   6, 7
electric field
– homogeneous ∼   SS166
– weak ∼   SS171
electron   SS125
– angular momentum   PE138–139
– Hamilton operator for two ∼s
   PE155
– in magnetic field
– – Hamilton operator   PE149
electrostatic interaction   SS125
energy conservation   SS173
energy eigenvalues
– continuum of ∼   190
– discrete ∼   190
energy spread   151, 153
entanglement   68
entire function   SS87
equation of motion
– Hamilton's ∼   SS46, PE32
– Heisenberg's ∼   2, 86, SS46, 48,
   PE31
– interaction picture   PE71
– Liouville's ∼   SS48, PE33
– Newton's ∼   2, 133
– Schrödinger's ∼   2, 85, SS44,
   PE31
– von Neumann's ∼   89, SS48
Esaki, Leo   200
Euler, Leonhard   32, SS151, PE42,
   168
Euler's beta function integral   PE42
Euler's factorial integral   SS151,
   PE168
Euler's identity   32
Euler–Lagrange equation   133, 135
even state   (see also odd state)
   159, 170, 187, 192, SS153

evolution operator   150, 156, SS43, PE45, 48, 81
- Born series   PE47
- dynamical variables   PE81
- group property   PE49
- Schrödinger equation   PE82
expectation value   53, SS23, 28
- of hermitian operator   SS22
- probabilities as ~s   53, SS27
expected value   53
exponential decay law   PE62

factorial
- Euler's integral   SS151, PE168
- Stirling's approximation   169
Fermi, Enrico   PE52, 158, 178
Fermi's golden rule   see golden rule
Fermi–Dirac statistics   PE158
fermions   PE158, 159, 164
Feynman, Richard. P.   SS139
flipper   16, 19, 21
- anti~   21
Fock, Vladimir A.   171, SS84, PE176
Fock states   171–175, SS84, 95, 97
- and coherent states   SS90–92
- completeness relation   SS88
- generating function   SS91
- momentum wave functions   SS92
- orthonormality   177, SS92
- position wave functions   175, 177, SS92
- two-dimensional oscillator   SS103
force   133, 155, SS144, PE41, 94
- ~s scatter   PE94
- constant ~   see constant force
- Lorentz ~   PE141
- of short range   183
- on magnetic moment   13
force-free motion   137, 141–154, 157–160, 164, PE143
- centrifugal potential   SS123
- completeness of energy eigenstates   157
- Hamilton operator   137, 141, SS49, 53, PE34

- Heisenberg equation   143, 144, SS53
- orthonormal states   159
- probability flux   PE91
- Schrödinger equation   141, 157, SS53, 54
- spread in momentum   144, SS58–63
- spread in position   144–148, SS58–63
- time transformation function   141–143, SS54, 72, PE34
- uncertainty ellipse   SS60–63
-- constant area   SS63
Fourier, Jean B. J.   123, SS3, PE24
Fourier integration   SS67
Fourier transformation   123, SS2, 16, 18, 51
Fourier's theorem   123
free particle   see force-free motion
frequency
- circular ~   166
- relative ~   53

g-factor   PE149
- anomalous ~   PE149
gauge function   PE41
Gauss, Karl F.   113, SS54, PE88
Gauss's theorem   PE88
gaussian integral   113, 131, 153, SS54
gaussian moment   SS3
generating function
- Fock states   SS91
- Hermite polynomials   176, SS92
- Laguerre polynomials   SS131
- Legendre polynomials   SS135
- spherical harmonics   SS136
generator   134, 135, SS101, PE36, 37, 39
- unitary transformation   SS101, 143
Gerlach, Walter   11, 14
Glauber, Roy J.   SS88
golden rule   PE52, 56
- applied to photon emission   PE59

– applied to scattering   PE97
Gordan, Paul A.   PE137
gradient   13, SS108, PE89
– spherical coordinates   SS121
Green, George   SS176, PE104
Green's function   PE104
– asymptotic form   PE107, 108
Green's operator   PE113
ground state   149, 195
– degenerate ∼s   PE78
– harmonic oscillator   179, SS81
– instantaneous ∼   PE76
– square-well potential   195
– two-electron atoms   PE167
ground-state energy   SS145
– constant restoring force   SS148
– delta potential   196
– harmonic oscillator   170
– lowest upper bound   SS148
– Rayleigh–Ritz estimate   SS146
– second-order correction   SS156
ground-state wave function
– harmonic oscillator   172
gyromagnetic ratio   see g-factor

half-transparent mirror   2, 4
Hamilton, William R.   84, SS43, 46,
    PE32
Hamilton function   135, SS43
Hamilton operator   84, 86, 88, 136,
    SS43, PE31
– arbitrary ∼   137
– atom-photon interaction   PE58,
    69
– bounded from below   149, 161
– center-of-mass motion   PE156
– charge in magnetic field
    PE141–143
– constant force   160, SS63, PE39
– driven two-level atom   PE69
– eigenbras   SS94
– eigenstates   156
– eigenvalues   156
– – degeneracy   157, PE144
– electron in magnetic field   PE149
– equivalent ∼s   87–88

– force-free motion   137, 141, SS49,
    53, PE34
– harmonic oscillator   165, 170, 171,
    SS72, 74, 78–80, 93
– hydrogenic atoms   SS125
– matrix representation   PE74
– metrical dimension   SS43
– photon   PE58, 69
– relative motion   PE156
– rotation   PE131
– spherical symmetry   SS120
– three-level atom   PE79
– time dependent ∼   87, SS45, PE32
– time-dependent force   SS67
– two electrons   PE155
– two-dimensional oscillator   SS98,
    118
– two-electron atoms   PE165
– two-level atom   PE57, 74
– typical form   137, 155, SS45, 172,
    PE88
– – virial theorem   SS142
Hamilton's equation of motion
    SS46, PE32
hard-sphere potential   PE112, 123
– bound states   PE126
– impenetrable sphere   PE126
– low-energy scattering   PE125
harmonic oscillator   165–180, SS72
– anharmonic perturbation   SS157
– eigenenergies   170
– energy scale   SS82
– ground state   179, SS81
– ground-state energy   170
– ground-state wave function   172
– Hamilton operator   165, 170, 171,
    SS72, 74, 78–80, 93
– – eigenkets   SS83
– – eigenvalues   SS83
– Heisenberg equation   165, SS72,
    78, 92
– ladder operators   173, SS83,
    PE58, 134
– length scale   166, SS79, 82
– momentum scale   SS79, 82
– momentum spread   180, SS84

– no-force limit SS77
– position spread 180, SS84
– Schrödinger equation 165, SS74
– time transformation function
  SS73, 75, 76, 93, 94, 96
– two-dimensional isotropic ∼ *see*
  two-dimensional oscillator
– virial theorem SS143
– wave functions 170, 172, 174, 175,
  177
– – orthonormality 177
– WKB approximation SS177
Hartree, Douglas R. PE176
Hartree–Fock equations PE176
Hausdorff, Felix SS40
Heaviside, Oliver SS181
Heaviside's step function SS181
Heisenberg, Werner 2, 86, 119, 127,
  SS34, 46, 61, PE22, 31
Heisenberg commutator 119, SS34,
  42, PE22, 25
– for vector components SS107
– invariance 119
Heisenberg equation (*see also*
  Heisenberg's equation of motion)
  86, SS46, 48, 80, PE31
– constant force 160, SS63, PE39
– force-free motion 143, 144, SS53
– formal solution 156
– general force 155
– harmonic oscillator 165, SS72, 78,
  92
– solving the ∼s 156
– special cases SS48
– time-dependent force SS67
Heisenberg picture PE49
Heisenberg's
– equation of motion (*see also*
  Heisenberg equation) 2, SS46, 48,
  PE31
– formulation of quantum mechanics
  119
helium SS171, PE165
helium ion SS125
Hellmann, Hans SS139

Hellmann–Feynman theorem
  SS139–141, 155, 165, PE150
Hermite, Charles 79, 169, SS22, 92,
  PE5
Hermite polynomials 169, 176–177
– differential equation 169
– generating function 176, SS92
– highest power 170
– orthogonality 176, 177
– Rodrigues formula 176
– symmetry 170
hermitian conjugate *see* adjoint
hermitian operator 78–80, SS22,
  PE5
– eigenvalues 79, PE5
– reality property SS22, PE5
Hilbert, David SS11, PE2
Hilbert space SS11, PE2
hydrogen atom (*see also* hydrogenic
  atoms) SS125
hydrogen ion PE165, 170, 171
hydrogenic atoms
– and two-dimensional oscillator
  SS127
– axis vector SS128
– Bohr energies SS139, 166
– Bohr shells SS139
– bound states SS128
– degeneracy SS128
– eigenstates SS128
– eigenvalues SS128
– Hamilton operator SS125
– mean distance SS141
– natural scales SS125
– radial wave functions SS134
– – orthonormality SS134
– scattering states SS128
– Schrödinger equation SS125
– total angular momentum PE138
– virial theorem SS143
– wave functions SS133–137

identity operator 40, SS14, PE3
– infinite matrix for ∼ 179
– square of ∼ 110
– square root of ∼ PE71, 73

– trace   58, 66
indeterminism   PE30
indistinguishable particles   PE155
– kets and bras   PE157
– permutation invariance of
  observables   PE155
– scattering of $\sim$ PE161–164
infinitesimal
– change of dynamics   PE34, 37
– changes of kets and bras   PE36
– endpoint variations   134
– path variations   135
– rotation   89, SS110
– time intervals   82
– time step   SS43
– transformation   SS101
– unitary transformation   83
– variation   SS75
– variations of an operator   PE41
inner product   SS4, 11, 12, PE2
integrating factor   161, SS65, 175
interaction picture   PE49, 71
interference   163, PE5
inverse
– unique $\sim$ 22

Jacobi, Karl G. J.   86, SS111
Jacobi identity   86, SS111
Jeffreys, Sir Harold   SS175

Kelvin, Lord $\sim$ see Thomson,
  William
Kepler, Johannes   SS128
Kepler ellipse   SS128
ket   35–38, SS10–16, PE1
– adjoint of a $\sim$ 37, PE1
– analog of column-type vector
  SS10
– basic $\sim$ 38, 71
– column for $\sim$ 99, 104
– eigen$\sim$ 46–50
– infinite column for $\sim$ 178
– metrical dimension   SS19
– normalization of $\sim$s   38, 71
– orthogonality of $\sim$s   38, 71
– phase arbitrariness   SS31

– physical $\sim$ SS11, 13
– reference $\sim$ 37
– row of $\sim$s   39, 55
ket-bra   38–41, SS24
– adjoint of a $\sim$ 40
kinetic energy   137, SS45, 117, 123
kinetic momentum   136
Kramers, Hendrik A.   SS175
Kronecker, Leopold   71, SS4, 165,
  PE3
Kronecker's delta symbol   see delta
  symbol

ladder operators   173
– angular momentum   SS113,
  PE132
– commutation relation   171, SS84
– differentiation with respect to $\sim$
  SS85
– eigenbras   SS86
– eigenkets   SS84
– eigenvalues   SS84
– harmonic oscillator   173, SS83,
  PE58, 134
– lowering operator   173
– orbital angular momentum
  SS113, 115
– raising operator   173
– two-dimensional oscillator   SS103,
  PE145
Lagrange, Joseph L.   133, PE174
Lagrange function   133
Lagrange parameter   PE174
Lagrange's variational principle   133
Laguerre, Edmond   SS119
Laguerre polynomials   SS131
– generating function   SS131
– Rodrigues formula   SS131
Lamb, Willis E.   PE64
Lamb shift   PE64
Langer, Rudolph E.   SS180
Langer's replacement   SS180
Laplace, Marquis de Pierre S.
  SS124, PE66
Laplace differential operator   SS124
Laplace transform   PE66

– inverse ~ PE67
– of convolution integral PE66
Larmor, Sir Joseph 92
Larmor precession 92
Legendre, Adrien M. SS135, PE119
Legendre function SS135
Legendre polynomials SS135, PE119
– generating function SS135
– orthonormality PE119
light quanta 2
Liouville, Joseph SS48, PE33
Liouville's equation of motion SS48
Lippmann, Bernard A. PE50
Lippmann–Schwinger equation PE84, 108, 112
– asymptotic form PE108
– Born approximation PE111
– exact solution PE116
– iteration PE50
– scattering operator PE50
lithium ion SS125, PE165
locality 6, 7
Lord Kelvin see Thomson, William
Lord Rayleigh see Strutt, John W.
Lord Rutherford see Rutherford, Ernest
Lorentz, Hendrik A. PE65, 141
Lorentz force PE141
Lorentz profile PE65

magnetic field 21
– charge in ~
– – circular motion PE146–148
– – Hamilton operator PE141–143
– – Lorentz force PE143
– – probability current PE143
– – velocity operator PE141
– homogeneous ~ 16, 30, 41, 42, 81, 92, PE143
– inhomogeneous ~ 11
– potential energy of magnetic moment in ~ 12
– vector potential PE141
magnetic interaction energy 86, 91, 101, PE149

magnetic moment 12, 92, PE129
– force on ~ 13
– potential energy of ~ in magnetic field 12
– rotating ~ PE129
– torque on ~ 12
many-electron atoms
– binding energy PE178
– outermost electrons PE180
– size PE179
matrices
– 2 × 2 ~ 17
– infinite ~ 177–180
– Pauli ~ see Pauli matrices
– projection ~ 29
– square ~ 17
– transformation ~ SS7
Maxwell, James C. 1
Maxwell's equations 2
mean value 53, 115, SS20
measurement
– disturbs the system 35
– equivalent ~s 82
– nonselective ~ 61–62
– result 53, PE2
– with many outcomes 70–75
mesa function SS27
metrical dimension
– bra and ket SS19, PE25
– Hamilton operator SS43
– Planck's constant SS19, 43
– wave function SS19
momentum 135
– canonical ~ PE142
– classical position-dependent ~ SS174
– kinetic ~ 136, PE142
momentum operator 117–119, 136, SS23
– differentiation with respect to ~ 155, SS35, 85, 108, PE25
– expectation value 124–125, SS59
– functions of ~ 122, SS23
– infinite matrix for ~ 179
– spread 144, 180, SS58
– vector operator SS107

momentum state   SS16
- completeness of ~s   120, SS16
- orthonormality of ~s   120, SS16
motion
- to the left   157, 189, SS174, PE92
- to the right   157, 189, SS174,
  PE92

Newton, Isaac   1, PE102
Newton's equation of motion   2, 133,
  165
Noether, Emmy   SS101
Noether's theorem   SS101
normalization   30
- force-free states   159
- state density   PE57
- statistical operator   SS31
- wave function   114, SS2, 3, 13

observables   PE1
- complementary ~   PE6, 12, 17, 30
- mutually exclusive ~   PE6
- undetermined ~   PE31
odd state   (see also even state)
  159, 170, 187, 192, SS153
operator   40, PE1
- adjoint of an ~   40
- antinormal ordering   SS88
- characteristic function   SS40
- equal ~ functions   73
- evolution ~   see evolution
  operator
- function of an ~   73, PE3, 4
-- unitary transformation of ~~
  77, SS38, PE20
-- varying a ~~   PE41–43
- hermitian ~   see hermitian
  operator
- identity ~   see identity operator
- infinite matrix for ~   177
- logarithm of an ~   PE43
- normal ~   72
- normal ordering   SS88
- ordered ~ function   SS34–37,
  PE13–16
- Pauli ~s   see Pauli operators

- Pauli vector ~   see Pauli vector
  operator
- projection ~   see projector
- reaction ~   see reaction operator
- scalar ~   SS112
- scattering ~   see scattering
  operator
- spectral decomposition   73, 83,
  SS22, PE3, 4
- spread   125
- statistical ~   see statistical
  operator
- unitary ~   see unitary operator
- vector ~   SS112, 116
optical theorem   PE114, 116
orbital angular momentum   SS105,
  109, PE130, 131
- commutators   SS109–113
- eigenkets   SS115
- eigenstates   SS113
- eigenvalues   SS105, 113, 115
- ladder operators   SS113, 115
- vector operator   SS109
-- cartesian components   SS109
ordered exponential   SS35
orthogonality   PE3
- of kets   38, 71
orthohelium   70
orthonormality   71, 83, SS6, PE3,
  23, 25, 28
- angular-momentum states   SS114
- azimuthal wave functions   SS133
- Fock states   177, SS92
- force-free states   159
- Legendre polynomials   PE119
- momentum states   120, SS16
- position states   111, SS13, 16
- radial wave functions   SS133, 134
- spherical harmonics   SS135
- time dependence   SS42
overidealization   114, SS11

partial waves
- for incoming plane wave   PE119
- for scattering amplitude   PE122
- for total wave   PE121

particle
- identical ∼s *see* indistinguishable particles
- indistinguishable ∼s *see* indistinguishable particles
Pauli, Wolfgang 43, SS116, PE129, 159
Pauli matrices 41–44, PE134
Pauli operators 41–44, 51
- functions of ∼∼ 44–45
- nonstandard matrices for ∼∼ 44, 56
- standard matrices for ∼∼ 43, 56
- trace of ∼∼ 58
Pauli vector operator 43, 57, SS116, PE129
- algebraic properties 43
- commutator of components 90
- component of ∼∼ 48
Peierls, Rudolf PE116
permutation
- cyclic ∼ PE7, 18
persistence probability PE54
perturbation theory 156, PE42
- Brillouin–Wigner *see* Brillouin–Wigner perturbation theory
- for degenerate states SS164–172
- Rayleigh–Schrödinger *see* Rayleigh–Schrödinger perturbation theory
phase factor 31, SS66
phase shift 187
phase space SS60
phase-space function SS33, 37
phase-space integral SS36, 37, 182
photoelectric effect 51
photon 2
- Hamilton operator PE58, 69
photon emission PE57–68
- golden rule PE59
- probability of no ∼ PE64
- Weisskopf–Wigner method PE60–65
photon mode PE59, 60
photon-pair source 4

Placzek, George PE116
Planck, Max K. E. L. 84, SS2, PE22
Planck's constant 84, PE22
- metrical dimension SS19, 43
Poisson, Siméon-Denis PE27
Poisson identity PE27
polar coordinates SS117
polarizability SS172
position operator 114–116, SS21
- delta function of ∼ 182
- differentiation with respect to ∼ 155, SS35, 85, 108, PE25
- expectation value 115, 116, SS59
- functions of ∼ 116, SS22
- infinite matrix for ∼ 179
- spread 144, 180, SS58
- vector operator SS107
position state SS16
- completeness of ∼s 110, SS16
- orthonormality of ∼s 111, SS13, 16
position-momentum correlation SS60, 64
potential energy SS45, 120
- localized ∼ PE91, 95
- separable ∼ PE96, 99
potential well 183
power series method 166–170
prediction *see* statistical prediction
principal quantum number SS127
principal value PE63
- model for ∼ PE68
probabilistic laws 4
probabilistic prediction *see* statistical prediction
probability 23–32, SS1, 13, 14, PE2, 87
- amplitude 23–32, 37–39, 109, SS1, 13, PE2
-- column of ∼s 95, 97
-- time dependent ∼∼ 93–97, 105
- as expectation value 53, SS27
- conditional ∼ SS2
- continuity equation PE88, 90
- current density PE87, 90, 110
-- charge in magnetic field PE143

– density   115, SS1, PE87
– flux of ∼   PE87
– for reflection   190, 198, 200
– for transmission   190, 198, 200
– fundamental symmetry   74, SS14, PE2
– local conservation law   PE88
– of no change   148–154
– – long times   149
– – short times   149
– transition ∼   *see* transition probability
probability operator   *see* statistical operator
product rule
– adjoint   SS19
– commutator   86, SS35
– transposition   SS5
projection   29
– matrices   29
– operator   *see* projector
projector   29, 45, 52, 115, PE8, 11
– on atomic state   PE57
– to an x state   182
property
– objective ∼   15, 65

quantum action principle   PE36–44

Rabi, Isidor I.   PE58
Rabi frequency   PE58, 59, 72
– modified ∼   PE73, 85
– time dependent ∼   PE69, 78
radial density   SS134
radial quantum number   SS119
radial Schrödinger equation   SS123, PE118
Rayleigh, Lord ∼   *see* Strutt, John W.
Rayleigh–Ritz method   SS145–153, PE167, 173
– best scale   SS148
– excited states   SS152–153
– scale-invariant version   SS150
– trial wave function   SS146

Rayleigh–Schrödinger perturbation theory   SS153–159, 162, 163
reaction operator   PE50
reflection
– ∼ symmetry   157
relative frequency   53
relative motion
– Hamilton operator   PE156
residue   186, PE106
Riemann, G. F. Bernhard   PE24
Ritz, Walther   SS145
Robertson, Howard P.   127
Rodrigues, Benjamin O.   176, SS131
Rodrigues formula
– Hermite polynomials   176
– Laguerre polynomials   SS131
Rohrer, Heinrich   200
rotation   SS8, 99, PE28
– consecutive ∼s   SS110–112
– Hamilton operator   PE131
– internal ∼   PE130
– orbital ∼   PE130
– rigid ∼   PE130
– unitary operator   SS102, 110
row   28
– eigen∼   99
– for bra   99, 104
– of coordinates   SS4
– of kets   39, 55
row-type vector   SS4
– analog of bra   SS11
Rutherford, Lord Ernest   PE102
Rutherford cross section   PE102
Ry   *see* Rydberg constant
Rydberg, Janne   SS125, PE166
Rydberg constant   SS125, PE166

s-wave scattering   PE123–127
– delta-shell potential   PE127
– hard-sphere potential   PE125
scalar product   (*see also* inner product)   SS4
scaling transformation   SS144
– and virial theorem   SS144
scattering   190, PE91–127
– Born series   PE114

– by localized potential   PE96
– cross section
–– Coulomb potential   PE102
–– golden-rule approximation   PE98
–– Rutherford ∼∼   PE102
–– separable potential   PE99
–– Yukawa potential   PE101
– deflection angle   PE100, 116, 119
– elastic ∼   PE98
– elastic potential ∼   PE98
– electron-electron ∼   PE161–163
– forward ∼   PE114
– golden rule   PE97
– in and out states   PE103
– incoming flux   PE97
– inelastic ∼   PE98
– interaction region   PE96, 107
– low-energy ∼   see s-wave
   scattering
– of s-waves   see s-wave scattering
– of indistinguishable particles
   PE161–164
– right-angle ∼   PE162
– separable potential   PE116
– spherically symmetry potential
   PE99
– transition matrix element   PE98
– transition operator   PE112–114
–– separable potential   PE117
scattering amplitude   PE109
– and scattering cross section
   PE111
– and scattering phases   PE122
– and transition operator   PE112
– partial waves   PE122
scattering cross section   PE97
– and scattering amplitude   PE111
– and scattering phases   PE122
scattering matrix   PE95
scattering operator   PE48, 83
– Born series   PE49
–– equation of motion   PE83
–– integral equation   PE50
–– Lippmann–Schwinger equation
   PE50
scattering phase   PE122

scattering states   190
– delta potential   186–190
– hydrogenic atoms   SS128
– square-well potential   197–201
Schrödinger, Erwin   2, 85, 118, 121,
   166, SS1, 44, 153, PE29, 31
Schrödinger equation   (see also
   Schrödinger's equation of motion)
   2, 85, SS44, 48, PE31
– driven two-level atom   PE70, 71
– evolution operator   PE82
– for bras   85, SS44, PE31
– for column of amplitudes   97
– for kets   85, SS44, PE31
– for momentum wave function
   137, 138
– for position wave function   137,
   138, SS45
– force-free motion   141, SS53, 54
– formal solution   SS94
– harmonic oscillator   SS74
– initial condition   137
– radial ∼   see radial Schrödinger
   equation
– solving the ∼   SS49
– time independent ∼   100, 103,
   SS174, PE92
–– constant force   160, 161
–– delta potential   184, 187
–– force-free motion   157
–– harmonic oscillator   165
–– hydrogenic atoms   SS125
–– spherical coordinates   SS123
–– square-well potential   197
–– two-dimensional oscillator
   SS118, 127
– time transformation function
   SS49, 50
– two-level atom   PE60
Schrödinger picture   PE49
Schrödinger's
– equation of motion   (see also
   Schrödinger equation)   85, SS44,
   PE31
– formulation of quantum mechanics
   118

Schur, Issai   PE16
Schur's lemma   PE16
Schwarz, K. Hermann A.   SS14
Schwinger, Julian   PE17, 37, 50, 178
Schwinger's quantum action principle
    (*see also* quantum action principle)
    PE37
Scott, J. M. C.   PE178
Scott correction   PE178
selection   15
– successive ~s   16, 19
selector   15, 18, 19
selfadjoint   *see* hermitian
short-range force   183
sign function   181, SS27
silver atom   11, 12, 52, 63, 89, 91,
    101, 128, PE129
single-photon counter   3
singlet   PE138, 159, 162
– projector on ~   PE161
Slater, John   PE160
Slater determinant   PE160, 176
solid angle   PE97
solid harmonics   SS136
spectral decomposition   73, 83,
    SS22, 23, PE3, 14, 25
speed   PE142
speed of light   PE141
spherical Bessel functions   PE119
– asymptotic form   PE120
spherical coordinates   48, SS120,
    PE99, 105
– gradient   SS121
– position vector   SS121
– Schrödinger equation   SS123
spherical harmonics   SS134, PE118
– generating function   SS136
– orthonormality   SS135
spherical wave
– incoming ~   PE121
– interferes with plane wave   PE114
– outgoing ~   PE109, 110, 121
spin   12, SS116, PE131
spin-orbit coupling   PE154
spin-statistics theorem   PE159
spread   125

– in energy   151, 153
– in momentum   144, SS58
– in position   144, SS58
– vanishing ~   152
spreading of the wave function
    145–148, SS58, 60, PE91
square-well potential   191–201
– attractive ~   197
– bound states   191–195
– count of bound states   SS183
– ground state   195
– reflection probability   198, 200
– repulsive ~   197
– scattering states   197–201
– Schrödinger equation   197
– transmission probability   198, 200
– tunneling   197–201
Stark, Johannes   SS170, PE154
Stark effect
– linear ~   SS170
– quadratic ~   SS172
state   (*see also* statistical operator)
    57
– bound ~s   *see* bound states
– coherent ~s   *see* coherent states
– even ~   *see* even state
– Fock ~s   *see* Fock states
– mixed ~   SS30, 32
– odd ~   *see* odd state
– of affairs   1, SS3
– of minimum uncertainty   129–131
– of the system   1, 2, SS3
– pure ~   SS31, 32
– reduction   68–70
– scattering ~s   *see* scattering states
– vector   36
state density   PE57
– normalization   PE57
state of affairs   SS10, 26
state operator   *see* statistical
    operator
stationary phase   163
statistical operator   56–61, 126,
    SS26, 29, PE30
– blend   61, SS30
– – as-if reality   61, SS30

– for atom pairs 64, 65
– inferred from data SS29
– mixture 61, SS30
–– many blends for one ∼ 61, SS30
– nature of the ∼ 69
– normalization SS31
– positivity SS30
– represents information 62, 70, SS29
– time dependence 88–89, SS48, PE32
– time-dependent force SS72
statistical prediction 4, 9, 69, SS2, 29
– verification of a ∼ 25, SS2
step function SS181
Stern, Otto 11, 14
Stern–Gerlach
– apparatus 24, 26, 51, 52, 70
–– generalization 70
– experiment 11–14, 31, 51, 128
– magnet 12, 14, 26, PE129
– measurement 23, 26, 31, 36, 154
– successive ∼ measurements 14–17
Stirling, James 169
Stirling's approximation 169
Strutt, John W. (Lord Rayleigh) SS136
surface element PE88
symmetry
– and degeneracy 157, SS99, 128
– reflection ∼ 157

Taylor, Brook 117
Taylor expansion 163
Taylor's theorem 117
Thomas, Llewellyn H. PE178
Thomas–Fermi energy PE178
Thomson, William (Lord Kelvin) SS134, 145, 153
three-level atom PE78
– Hamilton operator PE79
time dependence
– dynamical ∼ 86, SS48, 49, 97, PE32

– parametric ∼ 86, SS45, 48, 49, 97, PE32
time ordering PE51
– exponential function PE52
time transformation function 139–140, SS49–51, 95, PE33
– as a Fourier sum SS96
– constant force SS66, 67, PE40
– dependence on labels SS55
– force-free motion 141–143, SS54, 72, PE34
– harmonic oscillator SS73, 75, 76, 93, 94, 96
– initial condition 140, SS49–51
– Schrödinger equation SS49, 50
– time-dependent force SS69–71
– turning one into another 140, SS51
time-dependent force
– Hamilton operator SS67
– Heisenberg equation SS67
– spread in momentum SS69
– spread in position SS69
– statistical operator SS72
– time transformation function SS69–71
Tom and Jerry 63–70
torque on magnetic moment 12
trace 54–56, SS24, PE15
– as diagonal sum 55
– as sum of eigenvalues 56, PE74
– cyclic property 56, SS28
– linearity 54, SS25
– of ordered operator SS36
– of Pauli operators 58
– of the identity operator 58, 66
transformation function 120, 123, SS17, PE25
– time ∼ see time transformation function
transition PE52, 53
– frequency 100, 107, PE53, 54
– operator (see also scattering, transition operator) PE57, 112
– probability PE53, 54, 56
– rate PE54, 56, 57, 59, 97

transposition   SS4, 11
– of a product   SS5
trial wave function   SS146
triplet   PE138, 159, 162
– projector on ∼   PE161
tunnel diode   200
tunnel effect   200
tunnel transistor   200
tunneling microscope   200
tunneling probability   200
two-dimensional oscillator
   SS98–105, PE145
– and hydrogenic atoms   SS127
– degeneracy   SS99
– eigenstates   SS99, 102–105, 118
– Fock states   SS103
– Hamilton operator   SS98, 118
– ladder operators   SS103, PE145
– radial wave functions   SS133
– – orthonormality   SS133
– rotational symmetry   SS99
– Schrödinger equation   SS118, 127
– wave functions   SS129–133
two-electron atoms   PE165–175
– binding energy   PE170
– direct energy   PE172
– exchange energy   PE172
– ground state   PE167
– Hamilton operator   PE165
– interaction energy   PE168
– single-particle energy   PE168
two-level atom
– adiabatic evolution   PE74–78
– driven ∼   PE68–78
– – Hamilton operator   PE69
– – Schrödinger equation   PE70, 71
– frequency shift   PE64, 67
– Hamilton operator   PE57, 74
– instantaneous eigenstate   PE76,
   77
– periodic drive   PE72
– projector on atomic state   PE57
– resonant drive   PE71
– Schrödinger equation   PE60
– transition operator   PE57
– transition rate   PE59, 64, 67

uncertainty
– state of minimum ∼   129–131
uncertainty ellipse   SS60–63
– area   SS63
uncertainty principle   127
uncertainty relation   125–128
– Heisenberg's ∼   127
– – more stringent form   SS61
– physical content   128
– Robertson's ∼   127
unit
– macroscopic ∼s   84
– microscopic ∼s   84
unit matrix   29
unit vector   48
unitary operator   75–79, SS20, 23,
   PE5
– eigenvalues   78, PE5
– for cyclic permutations   PE7
– for shifts   PE28
– period   PE8
– rotation   SS102, 110
– transforms operator function   77,
   SS38, PE20
unitary transformation   SS42, 101
– generator   SS101, 143
uphill motion   162

vector   35, 36
– coordinates of a ∼   36
– state ∼   36
vector potential
– asymmetric choice   PE144
– symmetric choice   PE144
vector space   SS10
velocity   PE88, 141
– commutation relation   PE142
virial theorem   SS142, 144
– and scaling transformation   SS144
– harmonic oscillator   SS143
– hydrogenic atoms   SS143
von Neumann, John   89, SS48, PE33
von Neumann equation   89, SS48,
   PE33

wave function   109, SS1

– its sole role   SS2
– momentum ∼   123, SS2
–– metrical dimension   SS19
– normalization   114, SS2, 3, 13
– position ∼   123, SS2
–– metrical dimension   SS19
– spreading   *see* spreading of the
  wave function
– trial ∼   SS146
wave train   128
wave-particle duality   51
Weisskopf, Victor F.   PE60
Wentzel, Gregor   SS175
Wentzel–Kramers–Brillouin   *see*
  WKB
Weyl, Hermann K. H.   SS38, PE12,
  17
Weyl commutator   SS38, PE12
Wigner, Eugene P.   SS159, PE60,
  151
Wigner–Eckart theorem   PE151
WKB approximation   SS172–183
– harmonic oscillator   SS177
– reliability criterion   SS177
WKB quantization rule   SS178, 179,
  183
– in three dimensions   SS180
– Langer's replacement   SS180

Yukawa, Hideki   PE100
Yukawa potential   PE100
– double ∼   PE101
– scattering cross section   PE101

Zeeman effect   PE154
Zeeman, Pieter   PE154
Zeno effect   32–35, 154
Zeno of Elea   34